Selected Papers on Differential Equations and Analysis

American Mathematical Society

TRANSLATIONS

Series 2 • Volume 215

Selected Papers on Differential Equations and Analysis

American Mathematical Society
Providence, Rhode Island

2000 *Mathematics Subject Classification.* Primary 46L37, 81T05, 60J65, 31A05, 31B05, 31B25, 31C35, 58Jxx, 62D05, 35K57, 35R35, 42B25.

Library of Congress Cataloging-in-Publication Data

Selected papers on differential equations and analysis.
p. cm. (American Mathematical Society translations, ISSN 0065-9290 ; ser. 2, v. 215)
"[Translations of articles] originally published in the journal Sugaku"—Foreword.
Includes bibliographical references.
ISBN 0-8218-3927-6 (acid-free paper)
1. Differential equations. 2. Functional analysis. 3. Operator algebras. 4. Potential theory (Mathematics). I. Sugaku. II. Series.

QA3 .A572 ser. 2, vol. 215
[QA370]
510 s—dc22
[515′.35] 2005052407

∞ The paper used in this book is acid-free and falls within the guidelines
established to ensure permanence and durability.
Visit the AMS home page at http://www.ams.org/

10 9 8 7 6 5 4 3 2 1   10 09 08 07 06 05

Contents

Foreword vii

Regularity in free boundary problems
GEORG S. WEISS 1

Behavior of spatial critical points and zeros of solutions of diffusion equations
SHIGERU SAKAGUCHI 15

Martin boundary and boundary Harnack principle for non-smooth domains
HIROAKI AIKAWA 33

Singular integrals and Littlewood-Paley functions
SHUICHI SATO 57

Convergence of metric measure spaces and energy forms
ATSUSHI KASUE 79

Subfactor theory and its applications: Operator algebras and quantum field theory
YASUYUKI KAWAHIGASHI 97

Brownian motion and value distribution theory of holomorphic maps and harmonic maps
ATSUSHI ATSUJI 109

Statistical inference in two-stage sampling
MAKOTO AOSHIMA 125

Foreword

This is a collection of several papers originally published in the journal *Sūgaku*. Normally these translated articles would have appeared in the American Mathematical Society journal *Sugaku Expositions*. To speed up publication, the AMS decided to publish them as a volume in its series American Mathematical Society Translations Series 2.

The volume begins with the article "Regularity in free boundary problems" by Georg S. Weiss, devoted to partial differential equations with free boundary. The author describes three particular free boundary problems: the obstacle problem, the heat equation with strong absorption, and a singular limit problem in combustion theory. For each of these problems the author presents recent results related to regularity conditions of weak solutions of the corresponding equations and, in particular, to conditions ensuring the smoothness of the boundary between regions corresponding to different phases.

The article "Behavior of spatial critical points and zeros of solutions of diffusion equations" by Shigeru Sakaguchi is devoted to properties of solutions of Cauchy and Neumann-type problems for the heat equations with N spatial variables. Define the set of hot spots $M(t)$ of a solution $u(y,t)$ of such an equation by

$$M(t) = \{x \in \mathbb{R}^N : u(x,t) = \max_{y \in \mathbb{R}^N} u(y,t)\}.$$

The article presents results about the evolution of the set $M(t)$ as $t \to \infty$. In particular, the author describes the cases when the origin is a stationary hot spot of a solution of the Dirichlet problem in a bounded domain in $\mathbb{R}^N$.

The article "Martin boundary and boundary Harnack principle for non-smooth domains" by Hiroaki Aikawa presents results generalizing the classical Poisson integral representation of a harmonic function of the unit ball in the n-dimensional ball to domains with non-smooth boundaries. For such domains, the ordinary boundary of a smooth domain is replaced by the so-called minimal Martin boundary, and the classical Herglotz's theorem gives the *Martin integral representation*, which is an analog of the Poisson integral. The author studies a class of non-smooth domains for which the Martin boundary can be described explicitly and clarifies the relationship between the geometric structure of a domain and the properties of harmonic functions in it.

In the article "Singular integrals and Littlewood-Paley functions", Shuichi Sato studies oscillating singular integral operators with kernels satisfying certain growth and smoothness conditions and surveys some results on the boundedness of such operators in various L^p spaces.

In the article "Convergence of metric measure spaces and energy forms", Atsushi Kasue considers the Dirichlet problem for harmonic functions and studies convergence of the direct calculus of variations methods for obtaining solutions using the general convergence properties of Riemannian manifolds within the general theory of Hausdorff-Gromov distance between metric spaces. The author also studies the analogous convergence of Riemannian vector bundles.

The article "Subfactor theory and its applications: Operator algebras and quantum field theory" by Yasuyuki Kawahigashi provides a review of recent applications of the theory of operator algebras in geometry and physics. The author starts with the presentation of results by Alain Connes on classification of factors and on the use of operator algebras as one of the main ingredients of Connes's approach to noncommutative geometry. Then he describes the applications of operator algebras in the classification of knots (subfactor theory of V. F. R. Jones) and in quantum field theory.

In the article "Brownian motion and value distribution theory of holomorphic maps and harmonic maps", Atsushi Atsuji gives a survey of results relating the Brownian motion on a Riemannian manifold M to the function theory on M. These properties can be viewed as a generalization of the so-called conformal invariance of the Brownian motion. The article starts with a presentation of the Davis theorem about a typical homotopic class of a Brownian path on the plane. Then the author describes recent results on the relationship between Picard-type theorems on M and martingales, as well as relations between the Nevanlinna theory and stochastic calculus.

Finally, the article "Statistical inference in two-stage sampling" by Makoto Aoshima discusses the two-stage sampling method in hypothesis testing. The article starts by describing the general two-stage scheme of hypothesis testing and its asymptotic and statistical properties. Then the author continues by presenting examples of how such a scheme is used in practical problems of hypothesis testing.

American Mathematical Society

Amer. Math. Soc. Transl.
(2) Vol. **215**, 2005

Regularity in free boundary problems

Georg S. Weiss

A "free boundary problem" is a problem where one solves a given partial differential equation in a domain that is not known a priori, meaning that there is no way to know the domain *before* knowing the solution. The boundary of such a domain is called a "free boundary" (see Figure 1). To avoid an underdetermined problem-setting, it is necessary to add an additional boundary condition on the free boundary to the usual boundary condition. As the free boundary depends on the solution, the problem as a whole is going to be a nonlinear problem even if one starts with a linear differential equation in each phase.

For motivation let us consider the process of melting ice into water. Unless one does an experiment, one has in this case no means of knowing the position of the boundary between ice and water. Moreover the boundary is moving in time. We may assume that the temperature is in each phase (solid and liquid) a solution of the heat equation, but finding an adequate condition on the free boundary separating the two phases is not an easy task. If one assumes for example that the free boundary is the 0-level set of the temperature, one is thereby excluding phenomena like undercooling. In the case of the Stefan problem with Gibbs-Thomson law ([**28**]), it is possible to model undercooling, but the problem becomes harder to handle from the mathematical point of view since the boundary between ice and water is not a level set any more.

The industrial applications of free boundary problems are so numerous that a mere overview would be beyond the scope of this article. Here we confine ourselves to the explanation of a few examples in some detail. Solidification and melting processes as well as phase separation processes are often mentioned as classical applications of free boundary problems, but there are many other applications as various as fluid jets, tumor growth, Black-Scholes market models and others.

This article originally appeared in Japanese in Sūgaku 54 (3) (2002), 225–234.
2000 *Mathematics Subject Classification.* Primary 35R35.

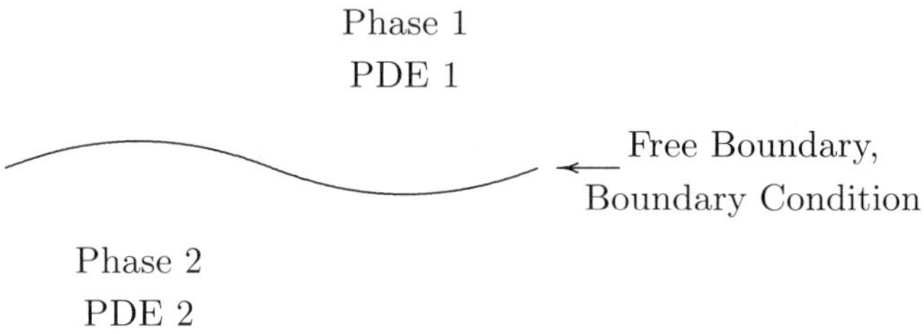

FIGURE 1. Free boundary problem

Let us mention that from the point of view of the engineer, classical solutions of free boundary problems – meaning solutions with free boundaries that are locally smooth surfaces such that the solutions can be extended to smooth functions on the closure of their respective phases – are to be desired. Reasons for this desire are that numerical analysis and the compactness of free boundaries which is necessary in inverse problems become much easier in the case of classical solutions. From the theoretical point of view, however, the situation is different: the existence of classical solutions of free boundary problems can rarely be shown in a direct way. If all goes well, one obtains a weak solution (e.g. a solution in the sense of distributions or a viscosity solution), and one has a priori no information on the regularity of that solution or the regularity of the free boundary. In the case of the classical Stefan problem, for example, a priori one does not even know whether the weak solution is continuous and whether the free boundary is a relatively closed set. Even when one has proved that the weak solution is continuous, the boundaries of the phases, $\partial\{u > 0\}$ and $\partial\{u < 0\}$, may have positive volume (think for example of a generalized Cantor set). Although the free boundary may be a level set, it is usually not possible to apply undergraduate tools like the implicit function theorem directly, as the solution or derivatives are discontinuous. In the example of the classical Stefan problem the first derivatives are discontinuous.

In these circumstances, regularity theory plays a vital role. The regularity theory of free boundary problems is – apart from the obvious relation to the regularity theory of nonlinear partial differential equations – deeply related to geometric measure theory and harmonic analysis. Much progress has been achieved here during the last few decades, the starting point being the "flatness implies regularity" results that have been proved in the time from the seventies to the early nineties. Roughly speaking, "flatness implies regularity" means that if the free boundary is in the neighborhood of a certain point sufficiently close to a plane, then it has to be a smooth surface in a certain smaller neighborhood of that point. Based on this success in proving regularity in the neighborhood of points that are in the above sense "flat", research is now focusing on the study of singular points (free boundary points that are not flat), free boundaries moving in time, and systems. It is to be hoped that the methods developed in free boundary problems will lead also to a framework that can be applied to other problems, like for example geometric partial differential equations and nonparametric minimal submanifolds.

In what follows we will discuss three selected free boundary problems: the obstacle problem, the heat equation with strong absorption, and a singular limit problem in combustion theory.

1. The obstacle problem

In order to motivate the term "obstacle problem", let us consider the process of spanning an elastic membrane over a given obstacle (see Figure 2).

In a simple model of the equilibrium situation we may assume the elastic membrane to be the graph of a function v and the obstacle to be the subgraph of a function g, where v minimizes the energy functional

$$\int_\Omega |\nabla v|^2$$

on the class $\{\phi \in H^{1,2}(\Omega) : \phi \geq g \text{ in } \Omega, \phi = h \text{ on } \partial\Omega\}$ of admissible functions. Let us assume here for the sake of simplicity that $\Delta g := \sum_{j=1}^n \frac{\partial^2}{\partial x_j^2} g < 0$. One can show that the minimizer v is in Ω a solution of the partial differential equation $\Delta v = \Delta g \chi_{\{v=g\}}$ in the sense of distributions, where χ_A is the characteristic function of the set A. Moreover, if we set $u := v - g$, then u is in Ω a non-negative solution of the partial differential equation

$$\Delta u = c\chi_{\{u>0\}}$$

in the sense of distributions; here $c = -\Delta g$ is a smooth positive coefficient.

Let us mention some mathematical results concerning the obstacle problem that are relevant in this context. By the Harnack inequality argument of H. Brézis and D. Kinderlehrer ([**7**]), the second derivatives of u are bounded. This regularity is sharp at each free boundary point, and u has quadratic growth at each free boundary point by the result of L.A. Caffarelli and N.M. Rivière in [**17**].

In the landmark paper [**10**], L.A. Caffarelli proved in 1977 that the free boundary is a smooth surface in a neighborhood of each free boundary point at which the coincidence set $\{u = 0\}$ has positive Lebesgue density. The result is based on blow-up arguments (arguments using blow-up sequences of the form $(\frac{u(x_m+r_m\cdot)}{r_m{}^2})_{m\in\mathbf{N}}$) and Harnack inequality arguments.

Step 1. By studying the blow-up limits (that is, the limits of the above blow-up sequences) and the second blow-up limits (the blow-up limits of the blow-up limits), one obtains the following: the free boundary is "flat" in a neighborhood $B_r(x_0)$ of each point x_0 at which the coincidence set $\{u = 0\}$ has positive Lebesgue density. This means that for each given $\epsilon > 0$ one can choose ρ small enough (depending on ϵ) so that $u = 0$ on $\{x \in B_\rho(x_0) : x \cdot \nu > \epsilon\rho\}$ and $u > 0$ on $\{x \in B_\rho(x_0) : x \cdot \nu < -\epsilon\rho\}$ for a certain direction ν (see Figure 3). The direction ν depends on ρ.

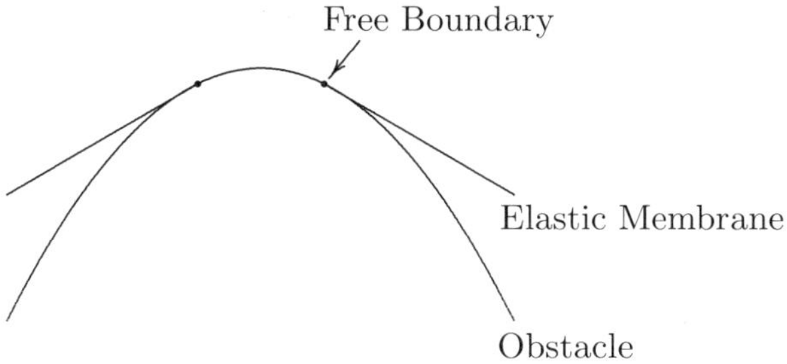

FIGURE 2. The obstacle problem in one dimension

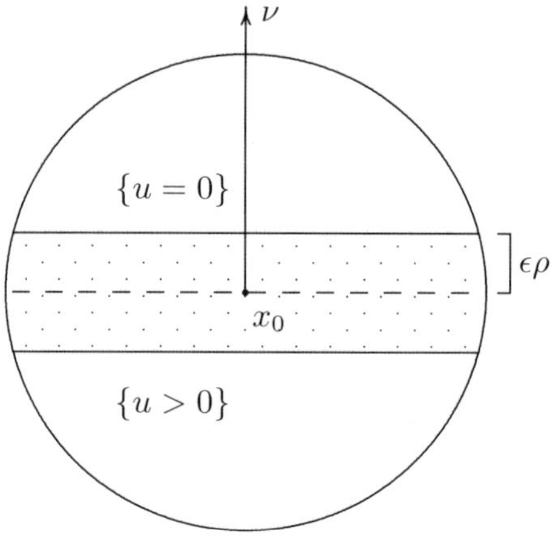

FIGURE 3. Flatness

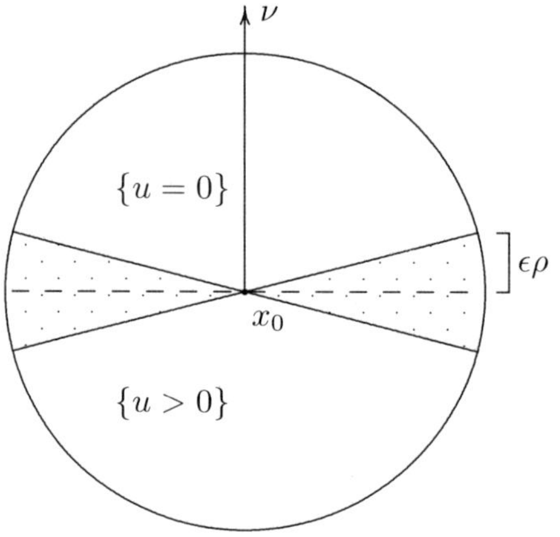

FIGURE 4. Cone-flatness

Step 2. Applying the Harnack inequality to second order partial derivatives of u, one obtains $\partial_{ee}u \geq -C|\log(\text{dist}(x, \{u = 0\}))|^{-\delta}$. Using this information, one can show "cone-flatness" of the free boundary, meaning that, choosing ρ small enough depending on ϵ, we get $u = 0$ in $\{x \in B_\rho(x_0) : x \cdot \nu > \epsilon|x - x_0|\}$ and $u > 0$ in $\{x \in B_\rho(x_0) : x \cdot \nu < -\epsilon|x - x_0|\}$ for a certain direction ν (see Figure 4). In contrast to flatness, which does not imply regularity of the free boundary, this cone-flatness directly implies the desired regularity of the free boundary.

Let us point out that for a general function c (a general obstacle) there may occur singularities, that is, points x_0 such that $\{u > 0\} \cap B_r(x_0)$ is not a subgraph of a C^1-function, no matter how small one chooses $r > 0$. For $c = 1$ the solutions $u(x) = \frac{1}{2}{x_1}^2$ and $v(x) = \frac{1}{2n}|x|^2$ give examples of singularities. In general it is possible to find singular sets of dimension k for each $k \in \{0, 1, \ldots, n-1\}$. Moreover, D. Schaeffer constructed an analytic c that gives rise to a cusp (see [**33**]). For c of class C^∞ the situation is even worse: for each relatively open subset U of $\{x_1 = 0\}$ there exist a c and boundary data such that $\{u > 0\} \cap \{x_1 = 0\}$ coincides with U ([**33**]). Consequently it is possible to construct a singular set whose boundary (relative to the free boundary) is a set of dimension $n-1$. This means that it is not possible to derive a partial regularity result for the free boundary or for the singular set.

What can be shown is the following. In a certain neighborhood of each singular point, the singular set is contained in a smooth surface. This was first proved in two dimensions by L.A. Caffarelli and N.M. Rivière via Harnack inequality arguments (see [**18**]). For analytic c, M. Sakai obtained a more precise result using methods of complex analysis ([**32**]): in the case of analytic c, the relative boundary of the singular set consists of isolated points. For the higher-dimensional case, L.A. Caffarelli proved in 1998 via the monotonicity formula in [**9**] that the singular set is contained in a smooth surface. I worked on the obstacle problem myself, when I developed my epiperimetric inequality approach to the regularity in free boundary problems with the obstacle problem as model problem (see [**36**]). The aim was – with applications to two-phase problems, systems and time-dependent problems in mind – to replace the existing methods based on Harnack inequality arguments by methods using energy functionals, which are given in many problems in a natural way. The epiperimetric inequality is an abstract, quantitative estimate in a given function space. I took the term "epiperimetric inequality" from the inequality derived by E. Reifenberg for the perimeter using completely different methods (see [**30**] and [**35**]). In the case of the obstacle problem one does not work with the perimeter, but with the energy functional

$$M(w) := \int_{B_1(0)} (|\nabla w|^2 + \max(w,0)) - 2\int_{\partial B_1(0)} w^2 d\mathcal{H}^{n-1}.$$

This energy satisfies for all non-negative functions $h \in H^{1,2}(B_1(0))$ that are homogeneous of degree 2 the following epiperimetric inequality: if

$$\|h - \frac{1}{4}\max(x\cdot\nu, 0)^2\|_{H^{1,2}(B_1(0))} \le \delta$$

for some $\nu \in B_1(0)$, then there exists a function $w \in H^{1,2}(B_1(0))$ that satisfies

$$M(w) \le (1-\kappa)M(h) + \kappa M(\frac{1}{4}\max(x\cdot\nu,0)^2)$$

as well as the boundary condition $w = h$ on $\partial B_1(0)$.

A possible interpretation of this theorem is that in a neighborhood of the "half-plane solutions" $\frac{1}{4}\max(x\cdot\nu,0)^2$ (solutions of the obstacle problem for the coefficient $c = \frac{1}{2}$), the energy satisfies a decay estimate. The proof does not use the Harnack inequality or the maximum principle, but is based on standard techniques in real and abstract analysis. Thus it is possible to think of extensions to other problems like systems etc. (actually the method has been extended to the time-dependent heat equation with strong absorption, which has a more complicated nonlinearity than the obstacle problem (see [**41**])). For the obstacle problem the technique has been successful in providing an alternative proof for the "flatness implies regularity" result. Moreover, deriving a similar epiperimetric inequality at singular solutions of the form $\frac{1}{4}|x\cdot\nu|^2$ and $\frac{1}{4n}|x|^2$, I recovered by completely different methods the result [**18**] of L.A. Caffarelli and N.M. Rivière – which I did not know at the time; ironically, although the author of [**32**] is working in Japan I heard of [**18**] and [**32**] for the first time when attending a meeting in Sweden.

Apart from the applications of the obstacle problem mentioned above, there are numerous real life applications in, say, filtration, elastic-plastic torsion, optimal control, mathematical finance, and so on (in [**23**] many applications are presented). Also, various two-phase problems related to the obstacle problem enjoy the interest of mathematicians studying the regularity of free boundaries (see for example [**13**], [**39**], [**34**] and [**5**]).

Let us also point out the relation between the obstacle problem and the Stefan problem mentioned above: if we assume that u is a non-negative solution of the classical Stefan problem

$$\partial_t(u + \chi_{\{u>0\}}) - \Delta u = 0$$

and that its time derivative satisfies $\partial_t u > 0$, then we obtain by integration in time that $v(t,x) := \int_0^t u(s,x)\,ds$ is a solution of the time-dependent obstacle problem

$$\partial_t v(t,x) - \Delta v(t,x) = -\chi_{\{v>0\}}(t,x) + f(x)\ .$$

In an open neighborhood of $((0,\infty)\times\mathbf{R}^n)\cap\{v=0\}$ the function $f=0$, so that one can treat this problem as a perturbation of the stationary obstacle problem (see [**10**]).

2. The heat equation with strong absorption

The heat equation with strong absorption

$$u \geq 0, \qquad \partial_t u - \Delta u = -u^\gamma,$$

where $\gamma \in (0,1)$ is a given parameter, is a problem of the same type as the time-dependent obstacle problem, but for physical reasons one is forced in this case to drop the assumption that $\partial_t u \geq 0$. This results in a big difference concerning the dynamics of the equation. Apart from singularities of the stationary problem, we encounter here singularities peculiar to the time-dependent problem, like for example extinction points, i.e. free boundary points in a spatial neighborhood of which the solution is 0.

The heat equation with strong absorption has been used by L. K. Martinson and by Ph. Rosenau and S. Kamin as a model for the transport of thermal energy in plasma (see [**29**] and [**31**]). Besides, it has been derived by C. Bandle and I. Stakgold in [**6**] as the asymptotic limit of a simple reaction-diffusion process.

In the case of one space dimension the solution's behavior near the extinction point has been extensively studied (see for example [**24**], [**20**], [**25**] and [**26**]). Let us take a closer look at two of these results: in [**24**], A. Friedman and M.A. Herrero analyzed the solution's spatial profile near extinction points. In [**20**], X.-Y. Chen, H. Matano and M. Mimura estimated the number of extinction points in the case of initial data with compact support. The authors also proved continuity of the set $\{u > 0\}$ in time with respect to the Hausdorff distance.

In the higher-dimensional case, H.W. Alt and D. Phillips obtained a result concerning the regularity of the solution and that of the free boundary ([**3**]): the proof of the solution's regularity, that is,

$$u^{\frac{1-\gamma}{2}} \in C^{0,1}_{loc},$$

relies – as in the case of the obstacle problem – on Harnack inequality arguments. The methods used in the proof of the regularity of the free boundary, however, have been adopted from a different problem, namely the problem with Bernoulli free boundary condition in [**1**]. We are going to discuss these methods in more detail in section 3 for the example of a Bernoulli type free boundary problem. As in the obstacle problem, the singular set of the heat equation with strong absorption can be of any integer dimension between 0 and $n-1$. The analysis of the structure of homogeneous solutions, however, is more difficult than that in the case of the obstacle problem. The reason is that the positive homogeneous solutions of degree 2 of the obstacle problem in $\mathbf{R}^n - \{0\}$ are second order polynomials, but a corresponding fact is not known for the equation $\Delta u = u^\gamma$.

Concerning the higher-dimensional case in the time-dependent problem, H. Choe and the author obtained in [**21**] regularity of the solution, finite propagation speed of the set $\{u > 0\}$, a Hausdorff measure estimate of the free boundary and the asymptotic behavior near "horizontal free boundary points". "Horizontal free boundary points" are free boundary points at which the behavior in time is dominant. Regularity of the solution means in this case that $u^{\frac{1-\gamma}{2}}$ is Lipschitz continuous with respect to the space variables and Hölder continuous with exponent $\frac{1}{2}$ with respect to the time variable. The proof relies, roughly speaking, on the fact that $|\partial_t(u^{1-\gamma})|^{\frac{1}{1-\gamma}}$ is a subsolution of the heat equation. The analysis of the asymptotic behavior is based on the following monotonicity formula ([**38**, Theorem 3.1]) obtained by the author:

$$\Psi^-_{(t_0,x_0)}(\sigma) - \Psi^-_{(t_0,x_0)}(\rho)$$

$$= \int_\rho^\sigma r^{-2\alpha-1} \int_{T_r^-(t_0)} \frac{1}{t_0-t} (\nabla u \cdot (x-x_0) - 2(t_0-t)\partial_t u - \alpha u)^2 G_{(t_0,x_0)}\, dr \geq 0.$$

Here

$$\alpha = \frac{2}{1-\gamma}, \qquad (t_0,x_0) \in (0,\infty)\times\mathbf{R}^n, \qquad T_r^-(t_0) = (t_0-4r^2, t_0-r^2)\times\mathbf{R}^n,$$

$$G_{(t_0,x_0)}(t,x) = 4\pi(t_0-t)|4\pi(t_0-t)|^{-\frac{n}{2}-1} \exp\left(-\frac{|x-x_0|^2}{4(t_0-t)}\right), \qquad 0<\rho<\sigma<\frac{\sqrt{t_0}}{2},$$

and

$$\Psi^-_{(t_0,x_0)}(r) = r^{-2(\alpha-1)-2} \int_{T_r^-(t_0)} \left(|\nabla u|^2 + \frac{2}{1+\gamma}\chi_{\{u>0\}} u^{1+\gamma}\right) G_{(t_0,x_0)}$$
$$- \frac{\alpha}{2} r^{-2\alpha} \int_{T_r^-(t_0)} \frac{1}{t_0-t} u^2\, G_{(t_0,x_0)}.$$

Concerning the regularity of the free boundary, I do not know whether the sophisticated technique used by H.W. Alt and D. Phillips in [**3**] can be extended to the time-dependent case. In [**41**], I applied the epiperimetric inequality approach developed earlier for the obstacle problem. Using the epiperimetric inequality, it was possible to show the existence of unique tangent cones and, eventually, to prove regularity: the free boundary can be decomposed into a "regular part" R and a "singular part" Σ, where $R = \{(t,x) \in ((0,\infty)\times\mathbf{R}^n) \cap \partial\{u>0\}$: at least one blow-up limit of u at (t,x) is a half-plane solution $\left(\frac{1-\gamma}{\sqrt{2(1+\gamma)}}\max(x\cdot\nu,0)\right)^{\frac{2}{1-\gamma}}\}$, the free boundary is a $\mathbf{C}^{\frac{1}{2},1+\mu}$-surface in an open neighborhood of R, and the singular part Σ is ignored by integration by parts in the set $\{u>0\}$. More precisely, for a.e. $t\in(0,\infty)$ and each $\zeta \in C_0^{0,1}(\mathbf{R}^n)$

$$\int_{\{u(t)>0\}} \partial_i\zeta = \int_{R\cap\{s=t\}} \zeta\nu_i \mathcal{H}^{n-1}.$$

On the other hand, the fact that the singular part is ignored by integration by parts does not necessarily mean that Σ is a small set: the stationary solution

$$\left(\frac{1-\gamma}{\sqrt{2(1+\gamma)}}|x_1|\right)^{\frac{2}{1-\gamma}}$$

presents here a counterexample (compare to the obstacle problem).

Higher regularity of R is an open problem.

3. A singular limit problem in combustion theory

In this section we consider the reaction-diffusion equation

$$\begin{aligned} &\partial_t u_\epsilon - \Delta u_\epsilon = -\beta_\epsilon(u_\epsilon), \\ &u_\epsilon \geq 0 \text{ in } (0,\infty)\times \mathbf{R}^n, \quad u_\epsilon(0,\cdot) = u_\epsilon^0 \text{ in } \mathbf{R}^n, \end{aligned} \tag{1}$$

and its singular limit as $\epsilon \to 0$; here $\epsilon \in (0,1), \beta_\epsilon(z) = \frac{1}{\epsilon}\beta(\frac{z}{\epsilon}), \beta \in C_0^1([0,1]), \beta > 0$ in $(0,1)$ and $\int \beta = \frac{1}{2}$. Formally each limit u with respect to a subsequence is a solution of the free boundary problem

$$\begin{aligned} &\partial_t u - \Delta u = 0 \text{ in } \{u>0\}\cap(0,\infty)\times\mathbf{R}^n, \\ &|\nabla u| = 1 \text{ on } \partial\{u>0\}\cap(0,\infty)\times\mathbf{R}^n. \end{aligned} \tag{2}$$

This limit problem has various applications: the stationary problem has been applied to fluid jets and cavities ([**2**]), electro-chemical machining ([**27**]), the construction of optimal heat conductors ([**22**]), etc. The singular limit problem (1) has been introduced as a model for the propagation of premixed flames in the case of high activation energy ([**8**], [**19**]). In this model $u = \lambda(T_c - T)$, where T_c is the flame temperature, which is assumed to be constant, T is the temperature outside the flame, and λ is a constant normalization factor.

Let us shortly summarize the known mathematical results, beginning with the limit problem. In the brilliant paper [**1**], H.W. Alt and L.A. Caffarelli showed existence of a stationary solution of (2) in the sense of distributions by minimizing the energy functional $\int(|\nabla u|^2 + \chi_{\{u>0\}})$. They furthermore proved regularity of the free boundary outside a set of $n-1$-dimensional Hausdorff measure 0. As the technique in their paper has since been applied to other problems like the heat equation with strong absorption (see [**3**]), we are going to discuss their methods in some detail. In this Bernoulli-type free boundary problem, too, the Harnack inequality plays a key role. It leads to Lipschitz continuity of the solution, and it is important when proving the following vital property: if the free boundary is flat on the side of $\{u=0\}$, then it must also be flat on the side of $\{u>0\}$. More precisely,

$$u = 0 \text{ in } B_r(x_0)\cap\{(x-x_0)\cdot\nu \geq \epsilon\} \Rightarrow u > 0 \text{ in } B_{cr}(x_0)\cap\{(x-x_0)\cdot\nu \leq -C\epsilon\}.$$

Here x_0 is a free boundary point, C and c are constants, and ν is a unit vector. The point is that the scaling of the left-hand side and that of the right-hand side in ϵ are the same.

Other valuable ideas are "partition of energy" and non-positive mean curvature of the free boundary of blow-up limits. "Partition of energy" means here that the Dirichlet part of the energy is asymptotically less than or equal to the volume part

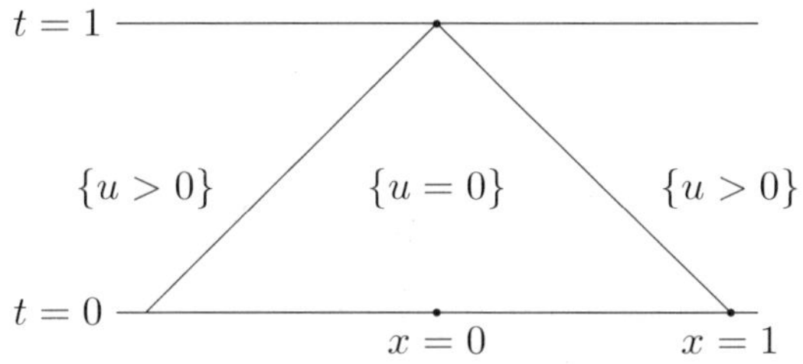

FIGURE 5. Colliding traveling waves

of the energy, i.e.

$$r^{-n} \int_{B_r(x_0)} |\nabla u|^2 \leq Cr^\alpha + r^{-n} \int_{B_r(x_0)} \chi_{\{u>0\}};$$

here, too, x_0 is assumed to be a free boundary point. In two space dimensions the inequality becomes an equality, that is, an "equipartition of energy". "Equipartition of energy" with respect to a different energy functional has later played a vital role in the context of mean curvature flow.

"Non-positive mean curvature of the free boundary of blow-up limits" refers to the fact that each limit u_0 of the blow-up sequence $\left(\frac{u(x_m+r_m\cdot)}{r_m}\right)_{m\in\mathbf{N}}$ satisfies for every open test set D the inequality $\mathcal{H}^{n-1}(D\cap\partial\{u_0 > 0\}) \leq \mathcal{H}^{n-1}(\partial D\cap\{u_0 > 0\})$. Let us point out, however, that the idea behind this fact has in [**1**] been applied to inhomogeneous blow-up limits and not to the homogeneous blow-up limits from above. Better regularity for inhomogeneous blow-up limits is used in [**1**] to derive a quantitative estimate, which eventually leads to regularity of the free boundary.

Existence of a classical solution in three dimensions is still an open problem, but it would follow from my result in [**40**] and [**37**], if non-existence of minimizing singular cones could be shown. Non-minimizing singular cones do in fact exist (see [**1**, Example 2.7]). Moreover, the solution of the stationary Dirichlet problem is in general not unique (see [**1**, Example 2.6]).

Let us return to the time-dependent problem (2). There is no uniqueness here, either. First, the positive solution of the heat equation is always another solution of the problem. Even if we exclude this trivial example of non-uniqueness, there are other counterexamples. Also, even if one starts with flawless initial data, a classical solution of (2) will in general develop singularities after a finite time-span. Consider, e.g., the example of two colliding traveling waves,

$$\begin{aligned} u(t,x) &= \chi_{\{x+t>1\}}(\exp(x+t-1)-1) \\ &\quad + \chi_{\{-x+t>1\}}(\exp(-x+t-1)-1) \quad \text{for } t \in [0,1) \end{aligned} \tag{3}$$

(cf. Figure 5).

Let us proceed with results concerning the singular limit problem (1). Concerning the stationary problem, H. Berestycki, L.A. Caffarelli and L. Nirenberg proved in [**4**] uniform estimates for the solution u_ϵ (concerning uniform estimates see also [**12**]). Assuming the existence of a minimal solution, the authors of [**4**] obtained further results. In [**19**], L.A. Caffarelli and J.L. Vazquez proved the corresponding uniform estimates in the case of time-dependence (see also [**14**]). Moreover they obtained a convergence result: if the initial data is strictly mean-concave in its

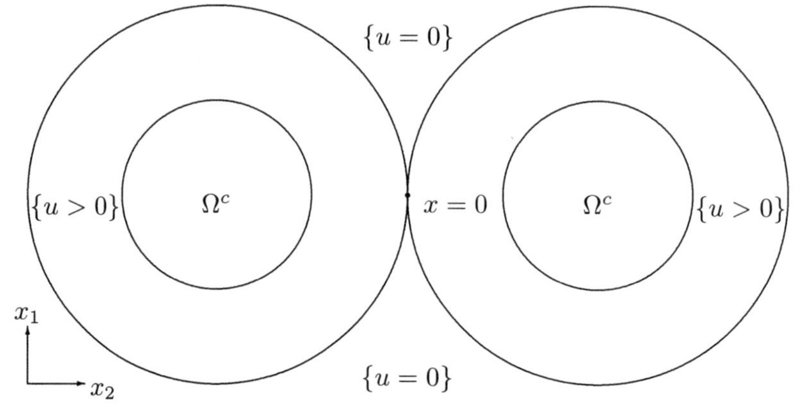

FIGURE 6. A singularity in the stationary limit problem

support, then each limit of solutions of problem (1) is a solution of problem (2) in the sense of distributions.

Let us also mention results concerning the corresponding two-phase problem. In [**15**] and [**16**], L.A. Caffarelli, C. Lederman and N. Wolanski show that if the limit function u satisfies $\{u = 0\}^\circ = \emptyset$, then it must be a sort of barrier solution.

Knowing the result of L.A. Caffarelli and J.L. Vazquez, one might conjecture that each limit of u_ϵ is always a solution in the sense of distributions. This, however, is not true: for arbitrary $\theta \in [0,1]$ there is u_ϵ converging to the limit $u(t,x) = \theta|x_1|$ which is not a solution of (2) in the sense of distributions. The same limit function u also appears as blow-up limit in the stationary problem and in the time-dependent problem (see Figures 6 and 5). For this reason, a class of solutions containing the function $u(t,x) = \theta|x_1|$ is desirable.

My contribution to this problem is as follows ([**42**]). First, each limit function is a solution in the sense of domain variations: u is smooth in $\{u > 0\}$ and satisfies for every $\xi \in C_0^{0,1}((0,\infty) \times \mathbf{R}^n; \mathbf{R}^n)$ the equation

$$(4) \quad \int_0^\infty \int_{\mathbf{R}^n} [-2\partial_t u \nabla u \cdot \xi + |\nabla u|^2 \operatorname{div} \xi - 2\nabla u D\xi \nabla u] = -\int_0^\infty \int_{R(t)} \xi \cdot \nu \, d\mathcal{H}^{n-1} \, dt.$$

Here

$$\begin{aligned} R(t) := \{x \in \partial\{u(t) > 0\} : \text{there is } \nu(t,x) \in \partial B_1(0) \text{ such that } u_r(s,y) \\ = \frac{u(t + r^2 s, x + ry)}{r} \to \max(-y \cdot \nu(t,x), 0) \\ \text{locally uniformly in } (s,y) \in \mathbf{R}^{n+1} \text{ as } r \to 0\} \end{aligned}$$

is for a.e. $t \in (0,\infty)$ a countably $n-1$-rectifiable subset of the free boundary (that is, a union of countably many C^1 manifolds and a set of $\mathcal{H}^{n-1}$-measure 0). Moreover, $R(t)$ is "almost" open relative to $\partial\{u(t) > 0\}$.

Furthermore, the free boundary can for a.e. $t \in (0,\infty)$ and up to a set of vanishing $\mathcal{H}^{n-1}$-measure be decomposed as $\partial\{u(t) > 0\} = R(t) \cup \Sigma_{**}(t) \cup \Sigma_z(t)$.

Here the non-degenerate singular set

$$\Sigma_{**}(t) := \{x \in \partial\{u(t) > 0\} \text{ :there is } \theta(t,x) \in (0,1] \text{ and } \xi(t,x) \in \partial B_1(0) \text{ such that } u_r(s,y) = \frac{u(t+r^2 s, x+ry)}{r} \to \theta(t,x)|y \cdot \xi(t,x)| \text{ locally uniformly in } (s,y) \in \mathbf{R}^{n+1} \text{ as } r \to 0\}$$

is a countably $n-1$-rectifiable subset of the free boundary, and the degenerate singular set

$$\Sigma_z(t) := \{x \in \partial\{u(t) > 0\} : r^{-n-2} \int_{Q_r(t,x)} |\nabla u|^2 \to 0 \text{ as } r \to 0\},$$

where $Q_r(t,x) = (t-r^2, t+r^2) \times B_r(x)$. That means that we can decompose the free boundary into a countably $n-1$-rectifiable part and the degenerate singular set. The crucial tool in the proof of this result is the following estimate of the mean frequency:

$$2\left(\int_{T_r^-(t)} \frac{1}{t-s} u^2\, G_{(t,x)}\right)^{-1} \int_{T_r^-(t)} |\nabla u|^2\, G_{(t,x)} \geq 1.$$

I was surprised when I obtained this estimate, which holds only on the singular set. The dimension of the degenerate singular set and the influence of the degenerate singular set on the limit equation pose open questions which are deeply related to harmonic analysis (see [**42**], [**43**]) and constitute a challenge to future research.

Thus the beauty of free boundary problems lies in the fact that it is not a closed field but takes its position in between partial differential equations, measure theory, geometry, harmonic analysis, etc. In the future we may expect fruitful joint research not only with the classical related fields but also with areas such as geometric equations, minimal submanifolds, and stochastic analysis.

References

[1] Alt, H.W. & Caffarelli, L.A., Existence and regularity for a minimum problem with free boundary, J. Reine Angew. Math. **325** (1981), 105-144.

[2] Alt, H.W., Caffarelli, L.A. & Friedman, A., Axially symmetric jet flows, Arch. Rational Mech. Anal. **81** (1983), 97-149.

[3] Alt, H.W. & Phillips, D., A free boundary problem for semilinear elliptic equations, J. Reine Angew. Math. **368** (1986), 63-107.

[4] Berestycki, H., Caffarelli, L.A. & Nirenberg L., Uniform estimates for regularization of free boundary problems, in Analysis and Partial Differential Equations, Marcel Dekker, New York, 1990, pp. 567–619.

[5] Apushkinskaya D.E., Shahgholian H. & Uraltseva N.N., On the global solutions of the parabolic obstacle problem, Algebra i Analiz **14** (2002), no. 1, 3–25; English transl., St. Petersburg Math. J. **14** (2003), 1–17.

[6] C. Bandle & I. Stakgold, The formation of the dead core in parabolic reaction-diffusion problems, Trans. Amer. Math. Soc. **286** (1984), 275-293.

[7] Brézis, H. & Kinderlehrer D., The smoothness of solutions to nonlinear variational inequalities, Indiana Univ. Math. J. **23** (1974), 831-844.

[8] Buckmaster, J.D. & Ludford, G.S.S., Theory of laminar flames, Cambridge University Press, Cambridge-New York, 1982.

[9] Caffarelli, L.A., A monotonicity formula for heat functions in disjoint domains, in Boundary value problems for partial differential equations and applications, Masson, Paris, 1993, pp. 53–60.

[10] Caffarelli L.A., The regularity of free boundaries in higher dimensions, Acta Math. **139** (1978), 155-184.

[11] Caffarelli, L.A., The obstacle problem revisited, J. Fourier Anal. Appl., **4** (1998), 383-402.
[12] Caffarelli, L.A., Uniform Lipschitz regularity of a singular perturbation problem, Differential Integral Equations **8** (1995), 1585-1590.
[13] Caffarelli, L.A., Karp, L. & Shahgholian, H., Regularity of a free boundary with application to the Pompeiu problem, Ann. Math. **151** (2000), 269-292.
[14] Caffarelli, L.A. & Kenig, C.E., Gradient estimates for variable coefficient parabolic equations and singular perturbation problems, Amer. J. Math. **120** (1998), 391-439.
[15] Caffarelli, L. A., Lederman, C. & Wolanski, N., Pointwise and viscosity solutions for the limit of a two phase parabolic singular perturbation problem, Indiana Univ. Math. J. **46** (1997), 719-740.
[16] Caffarelli, L. A., Lederman, C. & Wolanski, N., Uniform estimates and limits for a two phase parabolic singular perturbation problem, Indiana Univ. Math. J. **46** (1997), 453-489.
[17] Caffarelli, L.A. & Rivière, N.M., Smoothness and analyticity of free boundaries in variational inequalities, Ann. Scuola Norm. Super. Pisa Cl. Sci. IV **3** (1976), 289-310.
[18] Caffarelli, L.A. & Rivière, N.M., Asymptotic Behavior of Free Boundary Points at their Singular Points, Ann. Math. **106** (1977), 309-317.
[19] Caffarelli, L.A. & Vazquez, J.L., A free boundary problem for the heat equation arising in flame propagation. Trans. AMS **347** (1995), 411-441.
[20] Chen, X.-Y., Matano, H. & Mimura, M., Finite-point extinction and continuity of interfaces in a nonlinear diffusion equation with strong absorption, J. Reine Angew. Math. **459** (1995), 1-36.
[21] Choe H.J. & Weiss, G.S., A Semilinear Parabolic Equation with Free Boundary, Indiana Univ. Math. J. **53** (2003), 19–50.
[22] Flucher, M., An asymptotic formula for the minimal capacity among sets of equal area, Calc. Var. Partial Differential Equations **1** (1993), 71-86.
[23] Friedman, A., Variational principles and free-boundary problems. John Wiley & Sons, Inc., New York, 1982.
[24] Friedman, A. & Herrero, M.A., Extinction properties of semilinear heat equations with strong absorption. J. Math. Anal. Appl. **124** (1987), 530-546.
[25] Herrero, M.A. & Velazquez, J.J.L., On the dynamics of a semilinear heat equation with strong absorption, Comm. Partial Differential Equations **14** (1989), 1653-1715.
[26] Herrero, M.A. & Velazquez, J.J.L., Approaching an extinction point in one-dimensional semilinear heat equations with strong absorption, J. Math. Anal. Appl. **170** (1992), 353-381.
[27] Lacey, A.A. & Shillor, M., Electrochemical and electro-discharge machining with a threshold current, IMA J. Appl. Math. **39** (1987), 121-142.
[28] Luckhaus, S., Solutions for the two-phase Stefan problem with the Gibbs-Thomson law for the melting temperature, European J. Appl. Math. **1** (1990), 101-111.
[29] Martinson, L.K., The finite velocity of propagation of thermal perturbations in media with constant thermal conductivity, U.S.S.R. Comput. Math. and Math. Phys. **16** (1976), 141-149.
[30] Reifenberg, E.R., An epiperimetric inequality related to the analyticity of minimal surfaces, Ann. Math. **80** (1964) 1-14.
[31] Rosenau, Ph. & Kamin, S., Thermal waves in an absorbing and convecting medium, Physica D **8** (1983), 273-283.
[32] Sakai, M., Regularity of a boundary having a Schwarz function, Acta Math. **166** (1991), 263-297.
[33] Schaeffer, D.G., Some examples of singularities in a free boundary, Ann. Scuola Norm. Super. Pisa Cl. Sci. IV **4** (1977), 133-144.
[34] Shahgholian, H. & Uraltseva, N., Regularity properties of a free boundary near contact points with the fixed boundary, Duke Math. J. **116** (2003), 1–34.
[35] Taylor, J.E., Regularity of the singular sets of two-dimensional area-minimizing flat chains modulo 3 in $\mathbf{R}^3$, Invent. Math. **22** (1973), 119-159.
[36] Weiss, G.S., A homogeneity improvement approach to the obstacle problem, Invent. Math. **138** (1999), 23-50.
[37] Weiss, G.S., Partial regularity for weak solutions of an elliptic free boundary problem, Commun. Partial Differ. Equations **23** (1998), 439-457.
[38] Weiss, G.S., Self-similar Blow-up and Hausdorff-dimension Estimates for a Class of Parabolic Free Boundary Problems, SIAM J. Math. Anal. **30** (1999), 623-644.

[39] Weiss, G.S., An Obstacle-Problem-Like Equation with Two Phases: Pointwise Regularity of the Solution and an Estimate of the Hausdorff Dimension of the Free Boundary, Interfaces and Free Boundaries **3** (2001), 121-128.

[40] Weiss, G.S., Partial Regularity for a Minimum Problem with Free Boundary. J. Geom. Anal. **9** (1999), 317-326.

[41] Weiss, G.S., The Free Boundary of a Thermal Wave in a Strongly Absorbing Medium. J. Diff. Equations. **160** (2000), 357-388.

[42] Weiss, G.S., A Singular Limit arising in Combustion Theory: Fine Properties of the Free Boundary, Calc. Var. Partial Differential Equations **17** (2003), 311–340.

[43] Wolff, T., Plane harmonic measures live on sets of σ-finite length, Ark. Mat. **31** (1993), 137-172.

GRADUATE SCHOOL OF MATHEMATICAL SCIENCES, UNIVERSITY OF TOKYO, 3-8-1 KOMABA, MEGURO, TOKYO 153-8914, JAPAN

Translated by GEORG S. WEISS

Amer. Math. Soc. Transl.
(2) Vol. **215**, 2005

Behavior of spatial critical points and zeros of solutions of diffusion equations

Shigeru Sakaguchi

1. Introduction

Since solutions of partial differential equations are functions, it is natural to want to know their shapes. To know the shapes of graphs of functions, we begin by investigating their critical points. Let us consider the Cauchy problem for the heat equation as a plain example. In the N-dimensional Euclidean space $\mathbb{R}^N$ $(N \geqq 1)$, let $u = u(x,t)$ be the unique bounded solution of the problem for bounded initial data $u_0 \in L^\infty(\mathbb{R}^N)$. Then u satisfies

$$\partial_t u = \Delta u \quad \text{in } \mathbb{R}^N \times (0,\infty) \quad \text{and} \quad u(x,0) = u_0(x) \quad \text{in } \mathbb{R}^N. \tag{1.1}$$

Here, $x = (x_1, \dots, x_N) \in \mathbb{R}^N$, $\partial_t u = \frac{\partial u}{\partial t}$, and $\Delta u = \sum_{j=1}^N \frac{\partial^2 u}{\partial x_j^2}$. Of course, we have the integral representation of u via Gaussian kernel:

$$u(x,t) = (4\pi t)^{-\frac{N}{2}} \int_{\mathbb{R}^N} e^{-\frac{|x-y|^2}{4t}} u_0(y)\, dy. \tag{1.2}$$

Here $|x-y| = \left(\sum_{j=1}^N (x_j - y_j)^2\right)^{\frac{1}{2}}$, and $dy = dy_1 dy_2 \cdots dy_N$. Let us assume that u_0 satisfies the following:

$$u_0 \text{ is nonnegative and bounded, and } u_0 \not\equiv 0; \tag{1.3}$$

$$\text{the support of } u_0 \ (\mathit{supp}\ u_0) \text{ is a compact set.} \tag{1.4}$$

For each $t > 0$, define two sets $M(t)$ and $C(t)$ by

$$M(t) = \left\{x \in \mathbb{R}^N \ : \ u(x,t) = \max_{y \in \mathbb{R}^N} u(y,t)\right\}, \tag{1.5}$$

$$C(t) = \left\{x \in \mathbb{R}^N \ : \ \nabla u(x,t) = 0\right\}. \tag{1.6}$$

Here $\nabla u = (\frac{\partial u}{\partial x_1}, \dots, \frac{\partial u}{\partial x_N})$.

Note that $M(t) \subset C(t)$. $M(t)$ is called the set of hot spots, and $C(t)$ is called the set of spatial critical points. From the integral representation (1.2) of u, we obtain the following facts ([**ChaK**], [**JS**]).

(a) For each $t > 0$, $C(t)$ is contained in the closed convex hull of $\mathit{supp}\ u_0$.

This article originally appeared in Japanese in Sūgaku **54** (3) (2002), 249–264.
2000 *Mathematics Subject Classification.* Primary 35K57; Secondary 35B05.

(b) There exists a time $T > 0$ such that $C(t)$ consists of one point for each $t \geqq T$. Precisely, for each $t \geqq T$, there exists a point $x(t) \in \mathbb{R}^N$ satisfying $C(t) = M(t) = \{x(t)\}$.

(c) As $t \to \infty$, $x(t)$ tends to the Euclidean center of mass of u_0.

Let us consider an analogous problem for the initial-Dirichlet problem. We consider nonnegative initial data $u_0 \in L^\infty(\Omega)$ satisfying $u_0 \not\equiv 0$, where Ω is a bounded domain in $\mathbb{R}^N$ $(N \geqq 1)$. Let $u = u(x,t)$ be the unique solution of the following initial-Dirichlet problem for the heat equation:

$$\partial_t u = \Delta u \quad \text{in} \quad \Omega \times (0,\infty), \tag{1.7}$$

$$u = 0 \quad \text{on} \quad \partial\Omega \times (0,\infty), \tag{1.8}$$

$$u = u_0 \quad \text{on} \quad \Omega \times \{0\}. \tag{1.9}$$

Assume that Ω is a bounded convex domain. By replacing $\mathbb{R}^N$ with Ω in definitions (1.5) and (1.6), we define $M(t)$ and $C(t)$ similarly. Let $\varphi_1 = \varphi_1(x)$ be the positive first eigenfunction of $-\Delta$ satisfying $\int_\Omega \varphi_1^2(x)\,dx = 1$. Denoting by $\lambda_1 > 0$ the first eigenvalue yields

$$-\Delta\varphi_1 = \lambda_1\varphi_1 \ \text{ in } \Omega, \ \ \varphi_1 > 0 \ \text{ in } \Omega, \ \text{ and } \ \varphi_1 = 0 \ \text{ on } \partial\Omega. \tag{1.10}$$

A result of [**BL**] (see also [**Ko**]) shows that $\log\varphi_1$ is a concave function. Furthermore, a result of [**KoL**] shows that $\log\varphi_1$ is strictly concave, that is, the Hessian of $\log\varphi_1$ is negative definite. In particular, φ_1 attains its maximum at only one point, say x_∞. On the other hand, since the initial-Dirichlet problem (1.7)-(1.9) is solved by an eigenfunction expansion, by putting $c = \int_\Omega u_0(x)\varphi_1(x)\,dx$ we see that

$$e^{\lambda_1 t}u(x,t) \to c\varphi_1(x) \ \text{ in } L^2(\Omega) \text{ as } t \to \infty. \tag{1.11}$$

Furthermore, we see that the convergence in (1.11) is uniform on $\overline{\Omega}$, and by the regularity theory of solutions of parabolic partial differential equations any second-order partial derivative of $e^{\lambda_1 t}u$ with respect to space variables converges to that of $c\varphi_1$ uniformly on any compact set in Ω. In conclusion, there exist a time $T > 0$ and a point $x(t) \in \Omega$ $(t \geqq T)$ such that $M(t) = \{x(t)\}$ for each $t \geqq T$ and as $t \to \infty$ $x(t)$ tends to the point x_∞ at which φ_1 attains its maximum.

Next, let us assume that Ω is a bounded convex domain and $u_0 \equiv 1$. Then it follows from the maximum principle for the heat equation that $0 < u(x,t) < 1$ $(x \in \Omega,\ t > 0)$. Moreover, we see that, for each time $t > 0$, every isothermic surface of u is convex and smooth, and the set of hot spots $M(t)$ consists of one point. Precisely, a result of [**BL**] (see also [**Ko**]) shows that $\log u(x,t)$ is concave in x, which, together with the analyticity of u in x, implies that for each time $t > 0$ there exists a unique point $x(t) \in \Omega$ satisfying

$$C(t) = M(t) = \{x(t)\}. \tag{1.12}$$

Define the function $d = d(x)$ by

$$d(x) = \ \text{dist }(x, \partial\Omega) \ (= \inf_{y\in\partial\Omega} |x-y|) \quad (x \in \overline{\Omega}), \tag{1.13}$$

and put $I = \{x \in \Omega \ : \ d(x) = \max_{y\in\Omega} d(y)\}$. Then, we have

$$\text{dist }(x(t), I) \to 0 \ \text{ as } t \to +0, \tag{1.14}$$

since the function $-4t\log(1-u(x,t))$ attains its maximum at $x = x(t)$ for each $t > 0$ and a result of Varadhan [**V**] (see also [**FW**]) shows that

$$-4t\log(1-u(x,t)) \to d(x)^2 \quad \text{as } t \to +0 \text{ uniformly on } \overline{\Omega}. \tag{1.15}$$

In conclusion, the hot spot $x(t)$ starts from I and as $t \to \infty$ it tends to the point x_∞ at which the first eigenfunction φ_1 attains its maximum.

Concerning the behavior of the hot spot $x(t)$, let us give an example where $x(t)$ really moves. Let $N = 2$ and consider a semi-disk $\Omega = \{(x_1, x_2) \in \mathbb{R}^2 : x_1^2 + x_2^2 < 4,\ x_1 > 0\}$. Then $I = \{(1,0)\}$. Put $\Omega^\sharp = \{(x_1,x_2) \in \Omega : x_1 > 1\}$, and consider the function $v(x_1,x_2) = \varphi_1(2-x_1,x_2) - \varphi_1(x_1,x_2)$ for the first eigenfunction φ_1. Then v satisfies

$$-\Delta v = \lambda_1 v \quad \text{in } \Omega^\sharp \quad \text{and } v \geqq 0 \quad \text{on } \partial\Omega^\sharp.$$

Moreover, $v \not\equiv 0$, which, together with the monotonicity of the first eigenvalue with respect to the domain perturbation, implies that $v \geqq 0$ in $\Omega^\sharp$. (This follows also from the following argument. For the solution $u(x,t)$, consider the function $w(x_1,x_2,t) = u(2-x_1,x_2,t) - u(x_1,x_2,t)$. Then, w satisfies the heat equation in $\Omega^\sharp \times (0,\infty)$, $w \geqq 0$ on $\partial\Omega^\sharp \times (0,\infty)$, and $w(x,0) \equiv 0$. Hence it follows from the maximum principle for the heat equation that $w \geqq 0$. Therefore, the convergence (1.11) yields that $v \geqq 0$.) Furthermore, the minimum principle for superharmonic functions implies that $v > 0$ in $\Omega^\sharp$. Consequently, by using Hopf's boundary point lemma (see [**GT**], Lemma 3.4, p. 34) at the boundary point $(1,0) \in \partial\Omega^\sharp$, we obtain

$$0 < \frac{\partial v}{\partial x_1}(1,0) = -2\frac{\partial \varphi_1}{\partial x_1}(1,0).$$

This shows that $x_\infty \notin I(=\{(1,0)\})$, and really the hot spot $x(t)$ moves from the point $(1,0)$ to x_∞.

Also, if we replace $u_0 \equiv 1$ with $u_0(x) = \varphi_1(x)$, then the solution is given by $u(x,t) = e^{-\lambda_1 t}\varphi_1(x)$. In this case, $x(t) \equiv x_\infty$ and the hot spot never moves.

Concerning the initial-Neumann problem, there is Rauch's well-known conjecture which is related to the movement of the hot spots. Here, we only mention several references ([**At**], [**BnB**], [**BsB**], [**BW**], [**JN**]). In particular, there is a precise explanation of this conjecture in the introduction of Bañuelos and Burdzy [**BnB**]. In the papers [**At**], [**BnB**], [**BsB**], [**BW**] probabilistic techniques are used to prove the results, and only in [**JN**] are analytical techniques used.

The remainder of the present article is organized as follows. In Section 2, we mention similar results concerning solutions of nonlinear parabolic equations. In Sections 3 and 4, we explain two conjectures of Klamkin and report the author's recent results obtained by collaborating with R. Magnanini (Università di Firenze). In Section 5, we explain the balance law of the temperature around stationary critical points and stationary zeros of the heat flow, and we mention its role in proving the results in Sections 3 and 4. Finally in Section 6, we consider how a point is always a stationary critical point of solutions of the initial-boundary value problem for the heat equation for infinitely many initial data.

2. Nonlinear parabolic equations

When the space dimension $N = 1$ in particular, there are results which describe in detail the behavior of zeros and spatial critical points of solutions of parabolic equations (see [**Ma 1**], [**An**], [**Wa**], etc.). It is known that in general the number of zeros of each solution never increases with time. As an application of this property, for instance, Chen and Matano [**CheM**] studied the precise behavior of blow-up solutions and global in time solutions of initial-boundary value problems for semilinear heat equations. Also, Mizoguchi [**Mi**] studied the behavior of zeros of blow-up solutions and global in time solutions of the Cauchy problem for the heat equation and semilinear heat equations for initial data which change their sign at most finitely many times.

Aronson and Vazquez [**ArV**] studied the precise behavior of nonnegative solutions of the Cauchy problem for the one-dimensional porous medium equation $\partial_t u = \partial_x^2(u^m)$, where m is a number satisfying $m > 1$. This equation is a model which describes the diffusion of gas in a porous medium, and it has been studied for a long time. Let $u_0(x)$ be an initial data satisfying conditions (1.3) and (1.4). Since the diffusion coefficient mu^{m-1} vanishes on $\{u = 0\}$, this equation is degenerate parabolic. As a result of this degeneracy, it is known that this equation has the *property of finite speed of propagation*, that is, if the initial data has compact support, then for each $t > 0$ the function $u(\cdot, t)$ also does. In view of this property, in the descriptions (a), (b), and (c) in Section 1, it is natural for us to change the definition of the set $C(t)$ here as follows:

$$C(t) = \{x \in \mathbb{R} \; : \; \partial_x u(x,t) = 0 \text{ and } u(x,t) > 0\}. \tag{2.1}$$

It was shown in [**ArV**] that the pressure $v = (\frac{m}{m-1})u^{m-1}$ becomes strictly concave in x on the support of v after a finite time, and further, as $t \to \infty$, the hot spot $x(t)$ tends to the center of mass of the initial data u_0. Thus, (b) and (c) hold true. Also, Galaktionov and Vazquez [**GaV**] extended the results of [**ArV**] to the generalized porous medium equation $\partial_t u = \Delta\varphi(u)$.

In [**Sak 1**] and [**Sak 2**], the author considered the Cauchy problem for the following two nonlinear diffusion equations:

$$\partial_t u = \Delta(u^m) \quad \text{and} \quad \partial_t u = \text{ div } (|\nabla u|^{p-2}\nabla u).$$

The former is the porous medium equation, and the latter is called the p-Laplace evolution equation. Here, m and p are numbers satisfying $m > 0$ and $p > 1$ respectively, $|\nabla u| = (\sum_{j=1}^N (\frac{\partial u}{\partial x_j})^2)^{\frac{1}{2}}$, and we assume that the initial data u_0 satisfies (1.3) and (1.4). These equations equal the heat equation when $m = 1$ and $p = 2$, and they are degenerate or singular parabolic when $m \neq 1$ and $p \neq 2$. As a result of the degeneracy, when $m > 1$ and $p > 2$, these equations have the property of finite speed of propagation, and as a result of singularity, when $0 < m \leqq 1$ and $1 < p \leqq 2$, they have the *property of infinite speed of propagation*, that is, if the initial data u_0 is nonnegative and $u_0 \not\equiv 0$, then for each $t > 0$ the function $u(\cdot, t)$ is positive everywhere. The difficulty in dealing with the p-Laplace evolution equation is that the equation is degenerate or singular parabolic on the sets $M(t)$ and $C(t)$ which we have to consider. The results of [**Sak 1**] and [**Sak 2**] are the following:

(i) (a) holds true when $m > 0$, and (b) holds true when $m > \max\{\frac{N-2}{N}, 0\}$.
(ii) Both (a) and (b) hold true when $N = 1$ and $p > 1$.
(iii) (a) holds true when $N \geqq 2$ and $\frac{3}{2} < p < 2$.

It is conjectured that (a) holds true for any $p > 1$ when $N \geqq 2$. Basically (a) is proved by using Aleksandrov's reflection principle (see [**Ser**], [**GNN**], [**Ma 2**], etc.), which is the method comparing the solution with its reflection with respect to hyperplanes. To prove (b) of (ii), we introduced smooth approximate solutions. To the approximate solutions we applied the theory of Angenent [**An**], which gives precise information of zeros of solutions of one-dimensional parabolic equations, and the following argument of Ni and Sacks [**NS**]: *If a valley of the temperature continues to exist, then the temperature of the bottom of the valley never decreases, and this contradicts the eventual decay of the temperature. Therefore no valley of the temperature can continue to exist.* In addition to these we needed more arguments to complete the proof of (b) of (ii). To prove (b) of (i), we used a result of Friedman and Kamin [**FK**] concerning the asymptotic behavior of solutions as $t \to \infty$. It was shown in [**Sak 5**] that in (b) of (ii) the hot spot $x(t)$ is C^1 in t except at most one time.

3. Klamkin's conjecture — stationary hot spots

Let Ω be a bounded domain in $\mathbb{R}^N$ ($N \geqq 2$), and let $u = u(x,t)$ be the unique solution of the following initial-Dirichlet problem for the heat equation:

$$\partial_t u = \Delta u \quad \text{in} \quad \Omega \times (0, \infty), \tag{3.1}$$

$$u = 0 \quad \text{on} \quad \partial\Omega \times (0, \infty), \tag{3.2}$$

$$u = 1 \quad \text{on} \quad \Omega \times \{0\}. \tag{3.3}$$

There is a conjecture of Klamkin concerning the set of hot spots $M(t)$ of u. When Ω is a bounded convex domain in particular, as is mentioned in Section 1, for each time $t > 0$, every isothermic surface of u is convex and smooth, and the set of hot spots $M(t)$ consists of one point $x(t) \in \Omega$. Moreover (1.12) holds true. The conjecture of Klamkin [**Kl 2**] stated that in a bounded convex domain Ω containing the origin, if the origin is a stationary hot spot, that is, if $x(t) \equiv 0$, then the domain Ω must be centro-symmetric with respect to the origin. This was denied by Gulliver and Willms [**GW**] and Kawohl [**Ka**]. A typical counterexample is an equilateral triangle in the plane. After that Chamberland and Siegel [**CS**] posed the following conjecture.

CONJECTURE 3.1 (Chamberland and Siegel). *If the origin is a stationary hot spot in a bounded convex domain Ω, then the domain Ω is invariant under the action of an essential subgroup G of orthogonal transformations.*

A subgroup G is said to be *essential* if, for every $x \neq 0$, there exists an element $g \in G$ such that $gx \neq x$. Of course, the condition that the domain Ω is invariant under the action of an essential subgroup G of orthogonal transformations is a sufficient condition for the origin being a stationary hot spot. Indeed, if the domain Ω is invariant under the action of an essential subgroup G of orthogonal transformations, then by the unique solvability of the initial-Dirichlet problem (3.1)-(3.3) the solution u itself is invariant under the action of G. Namely, we have

$$u(x,t) \equiv u(gx,t) \quad (x \in \Omega,\ t > 0,\ g \in G).$$

This, together with the assumption that G is essential, implies that $\nabla u(0,t) = 0$ ($t > 0$), and then it follows from (1.12) that the origin is a stationary hot spot. There has been no result that would resolve Conjecture 3.1.

Our result concerning Conjecture 3.1 is

THEOREM 3.2 ([**MS 5**]). *When $N = 2$, the following hold true.*

(1) *If Ω is a triangle and the origin is a stationary hot spot, then Ω must be an equilateral triangle centered at the origin.*

(2) *If Ω is a convex quadrangle and the origin is a stationary hot spot, then Ω must be a parallelogram centered at the origin.*

In (1) of Theorem 3.2, G is the cyclic group generated by rotation through the angle $\frac{2\pi}{3}$, and in (2) $G = \{I, -I\}$ where I is the identity mapping. The proof is based on two ingredients; one is the balance law around stationary critical points of the heat flow (see [**MS 1**] and Section 5 of the present article) and the other is the asymptotic behavior as $t \to +0$ of solutions of the heat equation due to Varadhan [**V**]. Since the number 5 is prime, it is conjectured that if Ω is a convex pentagon and the origin is a stationary hot spot, then Ω must be a regular pentagon centered at the origin. This problem still remains to be solved.

4. Klamkin's conjecture—stationary isothermic surfaces

Let Ω be a bounded domain which is not necessarily convex, and let u be the unique solution of the initial-Dirichlet problem (3.1)-(3.3). Another conjecture of Klamkin [**Kl 1**] concerning isothermic surfaces of u stated that if the family of all the isothermic surfaces is invariant with time, then Ω must be a ball. (Of course, the temperature varies with time on each isothermic surface.) This conjecture was referred to by Zalcman [**Z**] from the point of view of inverse problems, and settled affirmatively by Alessandrini in [**A 1**] and [**A 2**]. In [**A 2**] Alessandrini considered the initial-Dirichlet problem where the initial data in (3.3) is replaced by any function $u_0 \in L^2(\Omega)$, and, with the help of Serrin's celebrated theorem (see [**Ser**], Theorem 2, pp. 311-312) that the existence of the solution of some elliptic boundary value problem with overdetermined boundary conditions implies that the domain is a ball, he showed the following: *If the family of all the isothermic surfaces of u is invariant with time, then either the initial data u_0 is an eigenfunction (that is, u is a separable solution) or Ω is a ball and u is radially symmetric in x with respect to the center of the ball Ω and monotone in the radial direction.* The analogous question for the initial-Neumann problem was examined and answered positively (see [**Sak 3**]) with the help of the classification theorem for isoparametric hypersurfaces in Euclidean space (see [**LC**], [**Seg**]), and the method used in [**Sak 3**] can be applied to give an alternative proof of the conjecture of Klamkin [**Kl 1**] concerning isothermic surfaces.

Here, in order to elucidate the relation between the conjecture of Klamkin [**Kl 1**] and isoparametric hypersurfaces, let us illustrate the method used in [**Sak 3**]. Fix a time $\tau > 0$ arbitrarily. For the unique solution u of the initial-Dirichlet problem (3.1)-(3.3), put $\psi(x) = u(x, \tau)$. Then, there exists an open subset $A \subset \Omega$ such that $\nabla\psi(x) \neq 0$ $(x \in A)$. Assume that in A all the isothermic surfaces are invariant with time. Then, as was observed by Alessandrini in [**A 1**] and [**A 2**], there exists a function $\mu = \mu(s, t)$ $(s \in \mathbb{R}, t \geqq 0)$ of two independent variables such that

$$u(x,t) = \mu(\psi(x), t) \quad (x \in A, t > 0).$$

Since $\nabla\psi(x) \neq 0$ $(x \in A)$, it follows from the implicit function theorem that μ is smooth. Hence by inserting μ into the heat equation we get

$$\mu_t = \mu_s \Delta\psi + \mu_{ss}|\nabla\psi|^2 \quad (x \in A, t > 0). \tag{4.1}$$

Differentiating this with respect to t yields

$$\mu_{tt} = \mu_{st} \Delta\psi + \mu_{sst}|\nabla\psi|^2 \quad (x \in A, t > 0). \tag{4.2}$$

Put $\mathfrak{D} = \mu_s \mu_{sst} - \mu_{ss}\mu_{st}$. When $\mathfrak{D} \equiv 0$, with the help of the analyticity of u in x, we see that u must be a separable solution. This contradicts the fact that u cannot be a separable solution under the initial condition (3.3). When $\mathfrak{D} \not\equiv 0$, we can solve a pair of equations (4.1)-(4.2) with respect to $|\nabla\psi|^2$ and $\Delta\psi$, and hence there exist two functions $f = f(s)$ and $g = g(s)$ of one real variable s such that

$$|\nabla\psi|^2 = f(\psi), \ \Delta\psi = g(\psi) \quad (x \in A).$$

This shows that the family of level surfaces of ψ, $\{x \in A \ : \ \psi(x) = s\}$ $(s \in \psi(A))$, consists of isoparametric hypersurfaces. Since all the isoparametric hypersurfaces in Euclidean space were completely classified by Levi-Civita and Segre (see [**LC**], [**Seg**]), this family must be either concentric spheres, parallel hyperplanes, or concentric spherical cylinders. Since ψ is positive and analytic in Ω and vanishes on $\partial\Omega$, the family of level surfaces of ψ in Ω must be compact and hence must be concentric spheres. Therefore Ω must be either a ball or an annulus. If we assume also that in the whole of Ω all the isothermic surfaces are invariant with time, then we can get the monotonicity in the radial direction of the radially symmetric function ψ (see [**Sak 3**], Lemma 2.2, pp. 230–231), and we see that Ω must be a ball.

An important observation is that, in order to prove the conjecture of Klamkin [**Kl 1**], both methods employed in [**A 1**]-[**A 2**] and [**Sak 3**] need to assume that infinitely many isothermic surfaces are invariant with time. Thus, the author posed the following conjecture at the conference titled "Nonlinear evolution equations and their applications" at Kyoto in October, 1997.

CONJECTURE 4.1 (Sakaguchi). *Let Ω be a bounded convex domain. Consider the unique solution u of the initial-Dirichlet problem* (3.1)-(3.3). *If one of the isothermic surfaces of u is invariant with time, then Ω must be a ball.*

The contrapositive of this conjecture tells us that if the bounded convex domain Ω is not a ball, then every isothermic surface of u must move.

Our answer to Conjecture 4.1 is

THEOREM 4.2 ([**MS 4**]). *Let Ω be a bounded domain in $\mathbb{R}^N$ $(N \geqq 2)$ satisfying the exterior sphere condition and suppose that D is a domain satisfying the interior cone condition, and such that $\overline{D} \subset \Omega$. Assume that the boundary ∂D is a stationary isothermic surface of the solution u of problem* (3.1)-(3.3), *that is,*

$$u(x,t) = a(t) \quad ((x,t) \in \partial D \times (0,\infty)) \tag{4.3}$$

for some function $a : (0,\infty) \to (0,\infty)$. Then Ω must be a ball.

Here, Ω is said to satisfy the *exterior sphere condition* if for every $y \in \partial\Omega$ there exists an open ball $B_r(z)$ such that $\overline{B_r(z)} \cap \overline{\Omega} = \{y\}$, where $B_r(z)$ denotes an open ball centered at $z \in \mathbb{R}^N$ and with radius $r > 0$. Also, D is said to satisfy the *interior cone condition* if for every $x \in \partial D$ there exists a finite right spherical cone K_x with vertex x such that $K_x \subset \overline{D}$ and $\overline{K_x} \cap \partial D = \{x\}$. These conditions describe a kind of regularity of $\partial\Omega$ and ∂D respectively. When Ω is a bounded convex domain in particular, the previous fact that every isothermic surface of u is convex and smooth, together with condition (4.3), implies that D is also convex and smooth, and hence all the assumptions of Theorem 4.2 are satisfied. Therefore Conjecture 4.1 holds true as a corollary of Theorem 4.2. We gave an outline of the proof of Theorem 4.2 in [**Sak 6**]. The point is obtaining the following lemma by combining the balance law of the temperature around stationary critical points and zeros of the heat flow (see [**MS 1**], [**MS 2**], and Section 5 of the present article) with the asymptotic behavior of the solution as $t \to +0$ due to Varadhan [**V**].

LEMMA 4.3. *Under the assumptions of Theorem 4.2 , there exists a positive number R satisfying*

$$\partial D = \{x \in \Omega \ : \ \mathrm{dist}(x, \partial\Omega) = R\}, \tag{4.4}$$

$$\prod_{j=1}^{N-1} (\frac{1}{R} - \kappa_j(y)) \equiv \text{ a positive constant } \quad (y \in \partial\Omega), \tag{4.5}$$

where $\kappa_1(y), \cdots, \kappa_{N-1}(y)$ are the principal curvatures of $\partial\Omega$ at y with respect to the interior normal direction to $\partial\Omega$ and satisfy $\kappa_j(y) < \frac{1}{R}$ $(j = 1, \cdots, N-1)$.

(4.4) is obtained by combining Varadhan's result (1.15) with assumption (4.3) of Theorem 4.2. We apply Aleksandrov's sphere theorem (see [**Alek**], p. 412) to Lemma 4.3. A special case of this sphere theorem is well-known as in the terms: *Soap bubbles are round* ([**R**]).

THEOREM 4.4 (Aleksandrov). *Let $N \geqq 3$, and let $\Phi = \Phi(\kappa_1, \cdots, \kappa_{N-1})$ be a C^1 function defined for $\kappa_1 \geqq \cdots \geqq \kappa_{N-1}$ and satisfying $\frac{\partial \Phi}{\partial \kappa_i} > 0$ $(i = 1, \cdots, N-1)$. Suppose that we have a closed C^2 hypersurface S embedded in $\mathbb{R}^N$. If on S the function Φ of its principal curvatures $\kappa_1, \cdots, \kappa_{N-1}$ is constant, then S must be a sphere.*

By using Lemma 4.3 and Theorem 4.4, let us prove Theorem 4.2. By Lemma 4.3, it suffices to consider the case where $N \geqq 3$. Letting the function Φ be set by

$$\Phi = \Phi(\kappa_1, \cdots, \kappa_{N-1}) = -\prod_{j=1}^{N-1} \left[\frac{1}{R} - \kappa_j\right] \tag{4.6}$$

yields that

$$\frac{\partial \Phi}{\partial \kappa_i} > 0 \quad (i = 1, \cdots, N-1), \text{ if } \max_{1 \leqq j \leqq N-1} \kappa_j < \frac{1}{R}.$$

Since by Lemma 4.3 condition (4.5) holds true, the function Φ is constant on $\partial\Omega$.

Therefore, by applying Theorem 4.4 to each connected component of $\partial\Omega$, we see that $\partial\Omega$ must be a sphere.

REMARK 4.5. The method of the proof of Theorem 4.4 is called *Aleksandrov's reflection principle* or the *method of moving planes*, which is based on the maximum principle for elliptic partial differential equations of second order. In fact, if we represent the surface S as the graph of a function g locally by using local coordinates, then the condition that $\Phi(\kappa_1, \ldots, \kappa_{N-1})$ is constant on S can be converted into the condition that g satisfies a second order elliptic partial differential equation, since $\frac{\partial \Phi}{\partial \kappa_i} > 0$ $(i = 1, \cdots, N-1)$. In the case where Φ is given by (4.6), this equation is a second order elliptic partial differential equation of Monge-Ampère type.

How about the case where the domain Ω is unbounded? By collaborating with R. Magnanini (Università di Firenze), the author recently obtained the following theorem by a proof almost similar to that of Theorem 4.2.

THEOREM 4.6 (Magnanini and Sakaguchi). *Let Ω be an exterior domain in $\mathbb{R}^N$ ($N \geqq 2$) satisfying the exterior sphere condition, and suppose that D is a domain satisfying the interior cone condition, and such that $\overline{D} \subset \Omega$. Assume that the boundary ∂D is a stationary isothermic surface of the solution u of problem* (3.1)-(3.3)*, that is, condition* (4.3) *holds true for some function $a : (0, \infty) \to (0, \infty)$. Then $\partial\Omega$ must be a sphere. That is, Ω must be the exterior of a ball.*

Since the boundary $\partial\Omega$ is compact, the proof of this theorem is based on the fact that we can use arguments similar to those used in the proof of Theorem 4.2 (see [**Sak 6**]).

Messoud A. Efendiev gave the following conjecture to the author at the conference titled "Nonlinear diffusive systems and related topics" at Kyoto in October, 2001.

CONJECTURE 4.7 (Efendiev). *Consider a domain Ω whose boundary $\partial\Omega$ is not compact. In particular, for a locally Lipschitz continuous function $\varphi = \varphi(x')$ ($x' = (x_1, \ldots, x_{N-1}) \in \mathbb{R}^{N-1}$), set*

$$\Omega = \{x \in \mathbb{R}^N \ : \ x_N > \varphi(x')\}.$$

If there exists a stationary isothermic surface of u of problem (3.1)-(3.3)*, then $\partial\Omega$ must be a hyperplane.*

Our answer to this conjecture is the following theorem.

THEOREM 4.8. *Let the function φ satisfy*

$$\nabla\varphi(x') = o(|x'|^{\frac{1}{2}}) \text{ near infinity,} \tag{4.7}$$

and suppose that Ω satisfies the uniform exterior sphere condition. Assume that there exists a domain D with $\overline{D} \subset \Omega$ such that ∂D is a stationary isothermic surface of the solution u of problem (3.1)-(3.3)*, that is, condition* (4.3) *holds true for some function $a : (0, \infty) \to (0, \infty)$. Furthermore, assume that the mean curvature of ∂D with respect to the exterior normal direction to ∂D is nonnegative everywhere. Then $\partial\Omega$ must be a hyperplane.*

Here, Ω is said to satisfy the *uniform exterior sphere condition* if there exists $r > 0$ such that for every $y \in \partial\Omega$ we have a ball $B_r(z)$ with $\overline{B_r(z)} \cap \overline{\Omega} = \{y\}$. Since the domain Ω is given by an unbounded domain above a locally Lipschitz graph, by using the maximum principle we see that $\frac{\partial u}{\partial x_N} > 0$ in $\Omega \times (0, \infty)$. Thus, in Theorem 4.8 the boundary ∂D is analytic by condition (4.3), and hence the mean curvature of ∂D is well-defined in Theorem 4.8.

With the help of a Bernstein type theorem due to Caffarelli, Nirenberg, and Spruck (see Theorem 2″ in [**CNS**]), we can prove this theorem. The details will be given in a forthcoming paper.

Every theorem (see Theorems 4.2, 4.6, and 4.8) concludes that $\partial\Omega$ must be an isoparametric hypersurface in $\mathbb{R}^N$. Thus, we may have a theorem which concludes that $\partial\Omega$ must be a spherical cylinder. Also, it is conjectured that we cannot have such a theorem where the surface concluded is not an isoparametric hypersurface. These are problems to be solved.

Note added in translation. When this translation was being made by the author himself, in the process he noticed that Theorem 4.8 needs one more assumption concerning the mean curvature of ∂D. This correction has been incorporated into the present translation.

5. Balance law and its role

In this section, we explain the balance law of the temperature around stationary critical points and zeros of the heat flow, which is mentioned in Sections 3 and 4.

THEOREM 5.1 (Balance law). *Let G be a domain in $\mathbb{R}^N$ $(N \geqq 2)$. Fix a point $x_0 \in G$, and set $d_* = \operatorname{dist}(x_0, \partial G)$. Suppose that $v = v(x,t)$ satisfies the heat equation in $G \times (0,\infty)$.*

Then the following hold:

(i) $v(x_0,t) = 0$ $(t > 0)$ *if and only if*

$$\int_{\partial B_r(x_0)} v(x,t)\, dS_x = 0 \quad ((r,t) \in (0,d_*) \times (0,\infty)), \tag{5.1}$$

where dS_x denotes the area element of the sphere $\partial B_r(x_0)$.

(ii) $\nabla v(x_0,t) = 0$ $(t > 0)$ *if and only if*

$$\int_{\partial B_r(x_0)} (x - x_0) v(x,t)\, dS_x = 0 \quad ((r,t) \in (0,d_*) \times (0,\infty)). \tag{5.2}$$

This balance law resembles the mean value theorem for harmonic functions, and it may be a kind of theorem in the function theory for functions satisfying the heat equation. (ii) was proved in [**MS 1**], and it was stated in [**MS 2**] that (i) can be proved similarly. It was shown in [**Sak 4**] that the balance law similar to (ii) holds true for the heat flow in N-dimensional sphere $\mathbb{S}^N$ and N-dimensional hyperbolic space $\mathbb{H}^N$. Also, in [**MS 4**], another proof of (i) and a shorter proof of (ii) with the aid of (i) were given. Here let us mention these proofs.

PROOF OF THEOREM 5.1. In (i), since sufficiency is obvious, let us show necessity. For $(r,t) \in [0,d_*) \times (0,\infty)$, define the function $p = p(r,t)$ by

$$p(r,t) = \int_{\mathbb{S}^{N-1}} v(x_0 + r\omega, t)\, d\omega, \tag{5.3}$$

where $\mathbb{S}^{N-1}$ is the unit sphere in $\mathbb{R}^N$ centered at the origin, $\omega = (\omega_1, \ldots, \omega_N) \in \mathbb{S}^{N-1}$ is a unit vector in $\mathbb{R}^N$, and $d\omega$ denotes the area element of $\mathbb{S}^{N-1}$. Then, p is analytic with respect to r on $[0,d_*)$ and satisfies

$$p_t = p_{rr} + \frac{N-1}{r} p_r \quad \text{in } (0,d_*) \times (0,\infty), \tag{5.4}$$

$$p(0,t) = p_r(0,t) = 0 \quad (t > 0). \tag{5.5}$$

Hence, (5.1) follows from the fact that

$$\frac{\partial^k p}{\partial r^k}(0,t) = 0 \quad (t > 0,\ k = 0,1,2,\ldots)$$

by induction on the integer k. (In [**Sak 4**] a similar argument was given in detail.)

Next, let us deduce (ii) from (i). Since every component of the vector $\nabla v(x,t)$ also satisfies the heat equation, $\nabla v(x_0,t) = 0$ $(t > 0)$ is equivalent to

$$\int_{\partial B_r(x_0)} \nabla v(x,t)\, dS_x = 0 \quad ((r,t) \in (0,d_*) \times (0,\infty)).$$

Furthermore, by integrating the latter formula with respect to r, we see that it is equivalent to

$$\int_{B_r(x_0)} \nabla v(x,t)\, dx = 0 \quad ((r,t) \in (0, d_*) \times (0,\infty)).$$

Therefore, by the divergence theorem, this formula is equivalent to (5.2). Thus (ii) is proved. □

Concerning the Cauchy problem for the heat equation (1.1), with the help of the integral representation of the solution (1.2), the initial data satisfies the balance law (5.2) with respect to a point $x_0 \in \mathbb{R}^N$ if and only if the point x_0 is a stationary critical point of the solution. Hence the following theorem holds:

THEOREM 5.2 ([**MS 1**]). *Fix a point $x_0 \in \mathbb{R}^N$. Let u be the unique solution of the Cauchy problem* (1.1) *for initial data $u_0 \in L^\infty(\mathbb{R}^N)$. Then, $\nabla u(x_0,t) = 0$ $(t > 0)$ if and only if*

$$\int_{\partial B_r(x_0)} (x - x_0) u_0(x)\, dS_x = 0 \quad (a.e.\ r > 0). \tag{5.6}$$

REMARK 5.3. When $N = 1$ in particular, it is easy to characterize stationary critical points of the heat flow. For simplicity, we consider the Cauchy problem (1.1). Let $x_0 \in \mathbb{R}$ be a stationary critical point of the solution u. Namely, suppose that $\partial_x u(x_0,t) = 0$ $(t > 0)$. Define the function $v = v(x,t)$ by

$$v(x,t) = \begin{cases} u(x,t) & \text{if } x < x_0, \\ u(2x_0 - x, t) & \text{if } x \geqq x_0. \end{cases}$$

Then, v also satisfies the heat equation in $\mathbb{R}\times(0,\infty)$, and $u = v$ in $(-\infty, x_0)\times(0,\infty)$. Hence, since both u and v are analytic in x, $u \equiv v$. Thus, we get the symmetry of u with respect to the point x_0:

$$u(x,t) = u(2x_0 - x, t) \quad ((x,t) \in \mathbb{R} \times (0,\infty)). \tag{5.7}$$

Therefore, the initial data u_0 is also symmetric with respect to x_0. Conversely, if we assume that the initial data u_0 is symmetric with respect to x_0, then by the unique solvability of the Cauchy problem (1.1) we get (5.7), and hence $x_0 \in \mathbb{R}$ is a stationary critical point of the solution u.

Next, let us explain the role of the balance law in the proofs of Theorem 3.2 and formula (4.5) of Lemma 4.3. For the unique solution u of the initial-Dirichlet problem (3.1)-(3.3), define the function $W = W(x,s)$ by

$$W(x,s) = s \int_0^\infty (1 - u(x,t))\, e^{-st} dt \quad ((x,s) \in \overline{\Omega} \times (0,\infty)). \tag{5.8}$$

Then, W satisfies the following:

$$\Delta W - s\, W = 0 \ \text{ in } \Omega \quad \text{and} \quad W = 1 \ \text{ on } \partial\Omega. \tag{5.9}$$

By a result of Varadhan [**V**] (see also [**EI**]), we have

$$-\frac{1}{\sqrt{s}} \log W(x,s) \to d(x) \ \text{ as } s \to \infty \text{ uniformly on } \overline{\Omega}, \tag{5.10}$$

where $d(x) =$ dist $(x, \partial\Omega)$ is the function defined by (1.13). In [**V**], Varadhan first showed (5.10) and then, by using it, he showed (1.15). Roughly speaking, (5.10) means that for sufficiently large $s > 0$ we may have an approximate formula

$$W(x, s) \sim e^{-\sqrt{s}d(x)}.$$

This formula, together with Laplace's method (see [**deB**], p. 71 or [**Shi**], pp. 100–101, and see also the stationary phase method [**Ev**], pp. 208–217), yields

THEOREM 5.4. *Let Ω be a bounded domain in $\mathbb{R}^N$ with piecewise C^2 boundary $\partial\Omega$, and denote by $\kappa_1, \ldots, \kappa_{N-1}$ the principal curvatures of the C^2 part of $\partial\Omega$ with respect to the interior normal direction to $\partial\Omega$. Assume that, for an open ball $B_r(x_0) \subset \Omega$, the set $\partial\Omega \cap \partial B_r(x_0)$ consists of a finite number of points $p_1, \ldots, p_K$, and $\kappa_j(p_k) < \frac{1}{r}$ $(j = 1, \ldots, N-1)$ for any $k = 1, \ldots, K$. Then, for any continuous function φ on $\mathbb{R}^N$, the following formula holds:*

(5.11)

$$\lim_{s \to \infty} s^{\frac{N-1}{4}} \int_{\partial B_r(x_0)} \varphi(x)\ W(x, s)\ dS_x = (2\pi)^{\frac{N-1}{2}} \sum_{k=1}^{K} \varphi(p_k) \left\{ \prod_{j=1}^{N-1} [\frac{1}{r} - \kappa_j(p_k)] \right\}^{-\frac{1}{2}}.$$

See [**MS 4**] for a detailed proof. In [**MS 5**], the case where the set $\partial\Omega \cap \partial B_r(x_0)$ consists of a finite number of submanifolds of the sphere $\partial B_r(x_0)$ was dealt with.

PROOF OF THEOREM 3.2 (1). Let Ω be a triangle in $\mathbb{R}^2$, and suppose that $x(t) \equiv 0$. Hence $\nabla u(0, t) = 0$ $(t > 0)$. Putting $v = 1 - u$ yields that $\nabla v(0, t) = 0$ $(t > 0)$ and v also satisfies the heat equation. Since (1.14) implies that $0 \in I$, by the definition of I the origin must be the inner center of the triangle Ω. By letting r be the radius of the inscribed circle, since v is continuous on $\overline{\Omega} \times (0, \infty)$, we have from the balance law (ii)

$$\int_{\partial B_r(0)} xv(x, t)\ dS_x = 0 \quad (t > 0).$$

Then the definition of W gives

$$\int_{\partial B_r(0)} xW(x, s)\ dS_x = 0 \quad (s > 0).$$

Put $\partial\Omega \cap \partial B_r(0) = \{p_1, p_2, p_3\}$ and observe that $\kappa_1(p_k) = 0$ $(k = 1, 2, 3)$. By considering continuous functions $\varphi(x) = x_j$ $(j = 1, 2)$ in Theorem 5.4, we conclude that $p_1 + p_2 + p_3 = 0$. This means that Ω is an equilateral triangle centered at the origin. □

REMARK 5.5. When Ω is a convex pentagon and $\partial\Omega \cap \partial B_r(0) = \{p_1, p_2, p_3, p_4, p_5\}$, the above argument yields that $p_1 + \cdots + p_5 = 0$. However, this equation does not guarantee that Ω is a regular pentagon centered at the origin (see [**MS 5**]).

Outline of the proof of (4.5). Let us begin by assuming that both $\partial\Omega$ and ∂D are smooth, (4.4) holds true, and that at any point $y \in \partial\Omega$ the principal curvatures $\kappa_1(y), \cdots, \kappa_{N-1}(y)$ of $\partial\Omega$ with respect to the interior normal direction to $\partial\Omega$ satisfy $\kappa_j(y) < \frac{1}{R}$ $(j = 1, \cdots, N-1)$. Then, observe that for any point $y \in \partial\Omega$, there exists a unique point $z = z(y) \in \partial D$ such that $\partial\Omega \cap \partial B_R(z(y)) = \{y\}$. Let p and q be two distinct points in $\partial\Omega$, and set $P = z(p)$, $Q = z(q)$. For $(x,t) \in B_R(0) \times (0,\infty)$, we define the function $v = v(x,t)$ by

$$v(x,t) = u(x+P,t) - u(x+Q,t). \tag{5.12}$$

Then, $v = v(x,t)$ satisfies the heat equation in $B_R(0) \times (0,\infty)$, and, by assumption (4.3) of Theorem 4.2,

$$v(0,t) = u(P,t) - u(Q,t) = 0 \quad (t > 0).$$

Since v is continuous on $\overline{B_R(0)} \times (0,\infty)$, by the balance law (i) we have

$$\int_{\partial B_R(0)} v(x,t)\, dS_x = 0 \quad (t > 0).$$

Hence

$$\int_{\partial B_R(P)} (1 - u(x,t))\, dS_x = \int_{\partial B_R(Q)} (1 - u(x,t))\, dS_x \quad (t > 0),$$

and by the definition of W

$$\int_{\partial B_R(P)} W(x,s)\, dS_x = \int_{\partial B_R(Q)} W(x,s)\, dS_x \quad (s > 0). \tag{5.13}$$

Multiply both sides of this equation by $s^{\frac{N-1}{4}}$ and consider the limits as $s \to \infty$. Since $\partial\Omega \cap \partial B_R(P) = \{p\}$ and $\partial\Omega \cap \partial B_R(Q) = \{q\}$, by considering a continuous function $\varphi(x) \equiv 1$ in Theorem 5.4 we conclude that

$$\prod_{j=1}^{N-1} \left[\frac{1}{R} - \kappa_j(p)\right] = \prod_{j=1}^{N-1} \left[\frac{1}{R} - \kappa_j(q)\right].$$

This is (4.5) itself. □

6. Problems for infinitely many initial data

We always considered the initial data $u_0 \equiv 1$ in Klamkin's conjectures. Here, we deal with spatial critical points as well as hot spots, and we consider how a point is always a stationary critical point of solutions of the initial-boundary value problem for infinitely many initial data. Let Ω be a smooth bounded domain in $\mathbb{R}^N$ $(N \geqq 2)$. Fix a number $\delta > 0$, and consider the domain Ω satisfying $B_\delta(0) \subset \Omega$, where $B_\delta(0)$ denotes an open ball centered at the origin and with radius $\delta > 0$. We consider the following initial-boundary value problem:

$$\partial_t u = \Delta u \quad \text{in } \Omega \times (0,\infty), \tag{6.1}$$

$$(1-\alpha)\frac{\partial u}{\partial \nu} + \alpha u = 0 \quad \text{on } \partial\Omega \times (0,\infty), \tag{6.2}$$

$$u(x,0) = u_0(x) \quad \text{in } \Omega, \tag{6.3}$$

where ν denotes the unit exterior normal vector to $\partial\Omega$, and α is a constant satisfying $0 \leqq \alpha \leqq 1$.

Let us define two families Φ and Φ_c of initial data u_0 by

$$\Phi = \{u_0 \in C_0^\infty(B_\delta(0)) \,:\, \int_{\mathbb{S}^{N-1}} \omega u_0(r\omega)\, d\omega = 0 \text{ for any } r \in [0,\delta)\},$$

$$\Phi_c = \{u_0 \in C_0^\infty(B_\delta(0)) \,:\, u_0(x) = u_0(-x) \text{ for any } x \in B_\delta(0)\},$$

where $C_0^\infty(B_\delta(0))$ denotes the set of C^∞ functions on $B_\delta(0)$ each of which vanishes identically on a neighborhood of the boundary $\partial B_\delta(0)$. Then $\Phi_c \subset \Phi$, and Φ is the family of initial data $u_0 \in C_0^\infty(B_\delta(0))$ satisfying the balance law (5.2) with respect to the origin. By Theorem 5.1 (ii), for initial data $u_0 \in C_0^\infty(B_\delta(0))$, the condition $u_0 \in \Phi$ is necessary for the origin to be a stationary critical point of the solution u of the initial-boundary value problem (6.1)-(6.3). Here, we have the following two theorems:

THEOREM 6.1 ([**MS 1**]). *For any initial data $u_0 \in \Phi$ the solution u of the initial-boundary value problem* (6.1)-(6.3) *always satisfies $\nabla u(0,t) = 0$ $(t > 0)$, if and only if $\Omega = B_R(0)$ for some $R > \delta$.*

THEOREM 6.2 ([**MS 2**]). *For any initial data $u_0 \in \Phi_c$ the solution u of the initial-boundary value problem* (6.1)-(6.3) *always satisfies $\nabla u(0,t) = 0$ $(t > 0)$, if and only if Ω is centro-symmetric with respect to the origin.*

We have similar theorems if we consider spatial zero points instead of spatial critical points, or if we consider the wave equation instead of the heat equation (see [**MS 2**]). Also, it was shown in [**Sak 4**] that similar theorems hold true for the heat flow in the N-dimensional sphere $\mathbb{S}^N$ and in N-dimensional hyperbolic space $\mathbb{H}^N$. (As far as Theorem 6.1 goes, only the case where $\alpha = 1$ was dealt with for the heat flow in $\mathbb{S}^N$ and $\mathbb{H}^N$.)

Consider a subgroup G of the orthogonal group $O(N)$, and define the family $\Phi(G)$ of initial data u_0 by

$$\Phi(G) = \{u_0 \in C_0^\infty(B_\delta(0)) \ : \ u_0(x) = u_0(gx) \ \text{ for any } (x,g) \in B_\delta(0) \times G\}.$$

Then, concerning the notion "essential" mentioned in Section 3, we have the following proposition.

PROPOSITION 6.3 ([**MS 3**]). *G is essential if and only if $\Phi(G) \subset \Phi$.*

Hereafter, we suppose that $N = 2$, Ω is a bounded and simply connected domain, and $\alpha = 1$ (that is, the initial-Dirichlet problem). For any integer $d \geqq 2$, let G_d be the cyclic group generated by rotation through an angle $\frac{2\pi}{d}$. Since G_d is essential, Proposition 6.3 implies that $\Phi(G_d) \subset \Phi$. Note that $\Phi_c = \Phi(G_2)$. Then the following theorem holds true:

THEOREM 6.4 ([**MS 3**]). *Let u be the solution of the initial-Dirichlet problem* (6.1)-(6.3) *with $\alpha = 1$. Then:*

(1) *Let $d = 2$ or $d = 3$. Then, $\nabla u(0,t) = 0$ $(t > 0,\ u_0 \in \Phi(G_d))$ if and only if Ω is invariant under the action of G_d.*
(2) *Let $d = 4$. Then, $\nabla u(0,t) = 0$ $(t > 0,\ u_0 \in \Phi(G_4))$ if and only if Ω is invariant under the action of G_2 (that is, Ω is centro-symmetric with respect to the origin).*

The proof was based on the method of complex analysis with the aid of the Riemann mapping theorem (see [**MS 3**] for details). The case where $d = 5$ is not known, as in the conjecture mentioned at the end of Section 3. By including this, we write one of the conjectures in [**MS 3**] regarded as a generalization of Theorem 6.4.

CONJECTURE 6.5 ([**MS 3**]). *Let $d \geqq 2$ be an arbitrary integer, and let u be the solution of the initial-Dirichlet problem* (6.1)-(6.3) *with $\alpha = 1$. Then, $\nabla u(0,t) = 0$ $(t > 0,\ u_0 \in \Phi(G_d))$ if and only if there exists a prime number p dividing d such that Ω is invariant under the action of G_p.*

Acknowledgement

The author would like to thank Professor Messoud A. Efendiev for giving him Conjecture 4.7. This work was partially supported by a Grant-in-Aid for Scientific Research (B) (♯ 12440042) from the Japan Society for the Promotion of Science.

References

[Alek] A. D. Aleksandrov, Uniqueness theorems for surfaces in the large V, Vestnik Leningrad Univ. 13, no. 19 (1958), 5–8. (English translation: Amer. Math. Soc. Transl., Ser. 2, 21 (1962), 412–415.)

[A 1] G. Alessandrini, Matzoh ball soup: a symmetry result for the heat equation, J. Anal. Math. 54 (1990), 229–236.

[A 2] G. Alessandrini, Characterizing spheres by functional relations on solutions of elliptic and parabolic equations, Applicable Anal. 40 (1991), 251–261.

[An] S. Angenent, The zero set of a solution of a parabolic equation, J. Reine Angew. Math. 390 (1988), 79–96.

[ArV] D. G. Aronson and J. L. Vazquez, Eventual C^∞−regularity and concavity for flows in one-dimensional porous media, Arch. Rational Mech. Anal. 99 (1987), 329–348.

[At] R. Atar, Invariant wedges for a two-point reflecting Brownian motion and the "hot spots" problem, Electron. J. Probab. 6 (2001), Paper No. 18, pages 1–19.

[BnB] R. Bañuelos and K. Burdzy, On the "hot spots" conjecture of J. Rauch, J. Funct. Anal. 164 (1999), 1–33.

[BsB] R. F. Bass and K. Burdzy, Fiber Brownian motion and the "hot spots" problem, Duke Math. J. 105 (2000), 25–58.

[BL] H. L. Brascamp and E. H. Lieb, On extensions of Brunn-Minkowski and Prékopa-Leindler theorems, including inequalities for log-concave functions, and with an application to the diffusion equation. J. Funct. Anal. 22 (1976), 366–389.

[BW] K. Burdzy and W. Werner, A counterexample to the "hot spots" conjecture, Ann. of Math. (2) 149 (1999), 309–317.

[CNS] L. Caffarelli, L. Nirenberg, and J. Spruck, On a form of Bernstein's theorem, in "Analyse Mathématique et Applications", 55–66, Gauthier-Villars, Paris, 1988.

[CS] M. Chamberland and D. Siegel, Convex domains with stationary hot spots. Math. Methods Appl. Sci. 20 (1997), 1163–1169.

[ChaK] I. Chavel and L. Karp, Movement of hot spots in Riemannian manifolds, J. Anal. Math. 55 (1990), 271–286.

[CheM] X.-Y. Chen and H. Matano, Convergence, asymptotic periodicity, and finite-point blow-up in one-dimensional semilinear heat equations, J. Differential Equations 78 (1989), 160–190.

[deB] N. G. de Bruijn, Asymptotic Methods in Analysis. Bibliotheca Mathematica. Vol. 4, North-Holland Publishing Co., Amsterdam; P. Noordhoff Ltd., Groningen; Interscience Publishers Inc., New York, 1958.

[Ev] L. C. Evans, Partial Differential Equations. American Mathematical Society, Providence, R.I. 1998.

[EI] L. C. Evans and H. Ishii, A PDE approach to some asymptotic problems concerning random differential equations with small noise intensities, Ann. Inst. H. Poincaré Anal. Non Linéaire 2 (1985), 1–20.

[FW] M .I. Freidlin and A .D. Wentzell, Random Perturbations of Dynamical Systems. Springer-Verlag, New York, 1984.

[FK] A. Friedman and S. Kamin, The asymptotic behavior of gas in an n-dimensional porous medium, Trans. Amer. Math. Soc. 262 (1980), 551–563.

[GaV] V. A. Galaktionov and J. L. Vazquez, Geometrical properties of the solutions of one-dimensional nonlinear parabolic equations, Math. Ann. 303 (1995), 741–769.

[GNN] B. Gidas, W. M. Ni and L. Nirenberg, Symmetry and related properties via the maximum principle, Comm. Math. Phys. 68 (1979), 209–243.

[GT] D. Gilbarg and N. S. Trudinger, Elliptic Partial Differential Equations of Second Order (Second Edition), Springer-Verlag, Berlin, 1983.

[GW] R. Gulliver and N. B. Willms, A conjectured heat flow problem, in Solutions, SIAM Rev. 37 (1995), 100–104.

[JN] D. Jerison and N. Nadirashvili, The "hot spots" conjecture for domains with two axes of symmetry, J. Amer. Math. Soc. 13 (2000), 741–772.

[JS] S. Jimbo and S. Sakaguchi, Movement of hot spots over unbounded domains in $\mathbb{R}^N$, J. Math. Anal. Appl. 182 (1994), 810–835.

[Ka] B. Kawohl, A conjectured heat flow problem, in Solutions, SIAM Rev. 37 (1995), 104–105.

[Kl 1] M. S. Klamkin, A physical characterization of a sphere, in Problems, SIAM Rev. 6 (1964), 61.

[Kl 2] M. S. Klamkin, A conjectured heat flow problem, in Problems, SIAM Rev. 36 (1994), 107.

[Ko] N. Korevaar, Convex solutions to nonlinear elliptic and parabolic boundary value problems. Indiana Univ. Math. J. 32 (1983), 603–614.

[KoL] N. Korevaar and J. Lewis, Convex solutions of certain elliptic equations have constant rank Hessians, Arch. Rational Mech. Anal. 97 (1987), 19–32.

[LC] T. Levi-Civita, Famiglie di superficie isoparametriche nell'ordinario spazio euclideo, Atti Accad. naz. Lincei. Rend. Cl. Sci. Fis. Mat. Natur. 26 (1937), 355–362.

[MS 1] R. Magnanini and S. Sakaguchi, The spatial critical points not moving along the heat flow, J. Anal. Math. 71 (1997), 237–261.

[MS 2] R. Magnanini and S. Sakaguchi, Spatial critical points not moving along the heat flow II : The centrosymmetric case, Math. Z. 230 (1999), 695–712, Corrigendum, 232 (1999), 389.

[MS 3] R. Magnanini and S. Sakaguchi, Stationary critical points of the heat flow in the plane, J. Anal. Math. 88 (2002), 383–396.

[MS 4] R. Magnanini and S. Sakaguchi, Matzoh ball soup: Heat conductors with a stationary isothermic surface, Ann. of Math. (2) 156 (2002), 931–946.

[MS 5] R. Magnanini and S. Sakaguchi, On heat conductors with a stationary hot spot, Ann. Mat. Pura Appl. 183 (2004), 1–23.

[Ma 1] H. Matano, Nonincrease of the lap-number of a solution for a one-dimensional semilinear parabolic equation, J. Fac. Sci. Univ. Tokyo, Sec. IA 29 (1982), 401–441.

[Ma 2] H. Matano, Asymptotic behaviour of the free boundaries arising in one phase Stefan problems in multi-dimensional spaces, Nonlinear Partial Differential Equations in Applied Science (Tokyo, 1982), North-Holland Math. Studies, vol. 81, North-Holland, Amsterdam, 1983, pp. 133–151.

[Mi] N. Mizoguchi, Asymptotic behavior of zeros of solutions for parabolic equations, J. Differential Equations 170 (2001), 51–67.

[NS] W. M. Ni and P. Sacks, The number of peaks of positive solutions of semilinear parabolic equations, SIAM J. Math. Anal. 16 (1985), 460–471.

[R] R. C. Reilly, Mean curvature, the Laplacian, and soap bubbles, Amer. Math. Monthly 89 (1982), 180–188, 197–198.

[Sak 1] S. Sakaguchi, The number of peaks of nonnegative solutions to some nonlinear degenerate parabolic equations, J. Math. Anal. Appl. 203 (1996), 78–103.

[Sak 2] S. Sakaguchi, Spatial critical points of nonnegative solutions of the evolution p-Laplacian equation : The fast diffusion case, Differential Integral Equations 10 (1997), 1049–1063.

[Sak 3] S. Sakaguchi, When are the spatial level surfaces of solutions of diffusion equations invariant with respect to the time variable?, J. Anal. Math. 78 (1999), 219–243.

[Sak 4] S. Sakaguchi, Stationary critical points of the heat flow in spaces of constant curvature, J. London Math. Soc. (2) 63 (2001), 400–412.

[Sak 5] S. Sakaguchi, Regularity of the interfaces with sign changes of solutions of the one-dimensional porous medium equation, J. Differential Equations 178 (2002), 1–59.

[Sak 6] S. Sakaguchi, Stationary isothermic surfaces of the heat flow, in "Nonlinear diffusive systems and related topics" (Japanese) (Kyoto, 2001), Sūrikaisekikenkyūsho Kōkyūroku No. 1258 (2002), 36–48.

[Seg] B. Segre, Famiglie di ipersuperficie isoparametriche negli spazi euclidei ad un qualunque numero di dimensioni, Atti Accad. Naz. Lincei. Rend. Cl. Sci. Fis. Mat. Natur. 27 (1938), 203–207.

[Ser] J. Serrin, A symmetry problem in potential theory, Arch. Rational Mech. Anal. 43 (1971), 304–318.

[Shi] T. Shiga, Lebesgue Sekibun kara kakuritsuron, (Japanese) Kyoritsukōza 21 seikinosūgaku 10 Kyoritsushuppan, 2000.

[V] S. R. S. Varadhan, On the behavior of the fundamental solution of the heat equation with variable coefficients, Comm. Pure Appl. Math. 20 (1967), 431–455.

[Wa] K. Watanabe, Remarques sur l'ensemble de zéro d'une solution d'une équation parabolique en dimension d'espace 1, J. Math. Soc. Japan 49 (1997), 817–832.

[Z] L. Zalcman, Some inverse problems of potential theory, Contemp. Math. 63 (1987), 337–350.

DEPARTMENT OF MATHEMATICAL SCIENCES, FACULTY OF SCIENCE, EHIME UNIVERSITY, 2-5 BUNKYO-CHO, MATSUYAMA-SHI, EHIME 790-8577 JAPAN

Translated by SHIGERU SAKAGUCHI

Amer. Math. Soc. Transl.
(2) Vol. **215**, 2005

Martin boundary and boundary Harnack principle for non-smooth domains

Hiroaki Aikawa

Dedicated to the Memory of Professor Matts Essén

1. Martin boundary

Harmonic functions, one of the most fundamental kinds of functions in analysis, have been studied for a long time and their precise properties are very well known. For example, let us assume that h is a harmonic function in the unit ball $B(0,1) \subset \mathbb{R}^n$, continuous up to its boundary. Then h can be represented as the Poisson integral:

$$\frac{1}{\sigma_n}\int_{S(0,1)} \frac{1-|x|^2}{|x-y|^n} h(y)d\sigma(y),$$

where $S(0,1)$ is the unit sphere and σ stands for the surface measure on $S(0,1)$, i.e., $(n-1)$-dimensional Hausdorff measure, and σ_n is the surface area of $S(0,1)$. The assumption that h is continuous up to the boundary is crucial. Without the assumption, the above integral representation need not hold and the boundary behavior of h may be very complicated. In order to study the global behavior of harmonic functions, we have to impose some hypotheses besides harmonicity. Among them, positivity may be the simplest. We have in fact the following theorem ([**49**]).

Herglotz Theorem. *Suppose h is a positive harmonic function on the unit ball $B(0,1)$. Then there exists a unique measure μ_h on the boundary $S(0,1)$ such that*

$$h(x) = \frac{1}{\sigma_n}\int_{S(0,1)} \frac{1-|x|^2}{|x-y|^n} d\mu_h(y).$$

This article originally appeared in Japanese in Sūgaku **55** (1) (2003), 1–19.

2000 *Mathematics Subject Classification.* Primary 31A05, 31B05, 31B25, 31C35.

Key words and phrases. Harmonic function, Green function, Martin boundary, Fatou theorem, Poisson integral, boundary Harnack principle.

This work was supported in part by Grant-in-Aid for (B) (2) (No. 15340046) and Exploratory Research (No. 13874023), Japan Society for the Promotion of Science.

If, furthermore, h is continuous up to the boundary, then μ_h coincides with the surface measure σ times the boundary values of h.

The Herglotz theorem yields the Fatou theorem concerning the boundary behavior of positive harmonic functions ([**43**]). For $\xi \in S(0,1)$ and $0 < \theta < \pi/2$ we let $\Gamma_\theta(\xi) = \{x \in B(0,1) : \langle \xi - x, \xi \rangle > |x - \xi| \cos\theta\}$ be the cone with vertex at ξ and aperture θ, where $\langle \xi - x, \xi \rangle$ stands for the inner product. We say that a function f on $B(0,1)$ has a *non-tangential* limit ℓ if

$$\lim_{x \to \xi, x \in \Gamma_\theta(\xi)} f(x) = \ell \quad \text{for every } \theta, \ 0 < \theta < \pi/2.$$

Fatou Theorem. *Every positive harmonic function on the unit ball $B(0,1)$ has a non-tangential limit at almost all boundary points on $S(0,1)$.*

It may be natural to extend the Herglotz and Fatou theorems to more general domains. Suppose D is a bounded smooth domain. Let $G(x,y)$ be the Green function for D, i.e., $-\Delta G(\cdot, y) = \delta_y$ in D in the distribution sense and $G(\cdot, y)$ vanishes on the boundary for each fixed $y \in D$. If a harmonic function h on D is continuous up to the boundary, then h is represented by the Poisson integral $\int_{\partial D} \frac{\partial}{\partial n_y} G(x,y) h(y) d\sigma(y)$. More generally, if h is positive on D, instead of being continuous on $\overline{D}$, then there exists a unique measure μ_h on ∂D such that $h(x) = \int_{\partial D} \frac{\partial}{\partial n_y} G(x,y) d\mu_h(y)$. This Poisson integral representation, not so explicit as for the ball, yields the Fatou theorem for positive harmonic function.

The situation is much more complicated for general domains. A Lipschitz domain may have wedges and its boundary may admit points having no normals. A fractal domain, such as a snowflake domain, may have boundary of Hausdorff dimension bigger than $n-1$ and the 'surface measure' may lose its meaning. Martin [**57**] established the integral representation of positive harmonic functions on general domains with the minimum requirement: the existence of the Green function. He introduced an ideal boundary, now called the *Martin boundary*, and succeeded in establishing the integral representation of positive harmonic functions. Let us give an outline of his idea. For the precise definition the reader is referred to the bibliography, e.g., [**23**].

Let h be a positive harmonic function on a general domain D with Green function $G(x,y)$. Take an exhaustion $\{D_j\}$ of D. Let

$$R_h^{D_j}(x) = \inf\{u(x) : \ u > 0 \text{ is superharmonic on } D \text{ with } u \geq h \text{ on } D_j\}.$$

The lower regularization $\widehat{R}_h^{D_j}$ of $R_h^{D_j}$ is a superharmonic function on D. This is called the *balayage* or regularized reduced function of h to D_j. By definition $\widehat{R}_h^{D_j}$ vanishes on the boundary and is harmonic in $D \setminus \partial D_j$, so that the balayage is a Green potential $\int G(\cdot, y) d\nu_j(y)$ of some measure ν_j on ∂D_j. Let us fix a point $x_0 \in D$. Since $\widehat{R}_h^{D_j} = h$ on D_j, we obtain an integral representation

$$h(x) = \int G(x,y) d\nu_j(y) = \int \frac{G(x,y)}{G(x_0,y)} G(x_0,y) d\nu_j(y) \quad \text{for } x \in D_j.$$

Let us consider a sequence of measures $d\mu_j(y) = G(x_0, y)d\nu_j(y)$. Substituting $x = x_0$ in the above integral representation, we observe that the total mass of μ_j is identically equal to $h(x_0)$. Thus, a subsequence has a weak limit μ_h, which gives an integral representation

$$h(x) = \int \frac{G(x,y)}{G(x_0,y)} d\mu_h(y) \quad \text{for all } x \in D.$$

However, this is not sufficient. The ratio of the Green functions $G(x,y)/G(x_0,y)$ may not extend continuously to the boundary. We, thus, consider the minimum ideal boundary Δ where the ratio extends continuously. Then the extension $K(x,y)$ of $G(x,y)/G(x_0,y)$ to $D \times \Delta$ gives the integral representation

$$h(x) = \int_\Delta K(x,y)d\mu_h(y).$$

The ideal boundary Δ and the kernel $K(x,y)$ are called the *Martin boundary* and the *Martin kernel.* Note that $K(x,y)$ is a positive harmonic function on D with $K(x_0, y) = 1$. The Martin kernel can be identified with a Martin boundary point. Unfortunately, the measure μ_h in the above integral representation need not be unique. This leads to a minimal boundary point, or an essential point, among Martin boundary points. In general, a positive harmonic function u is called minimal if every non-negative harmonic function majorized by u coincides with a constant multiple of u. If the Martin kernel $K(\cdot, y)$ is minimal, then y is called a minimal Martin boundary point. The set of all minimal Martin boundary points is called the *minimal Martin boundary* and is denoted by Δ_1. The remaining boundary is called the non-minimal Martin boundary and is denoted by Δ_0. Herglotz's theorem is now generalized as follows:

Martin Integral Representation. *For each positive harmonic function h on D there exists a unique measure μ_h on Δ_1 such that*

$$h(x) = \int_{\Delta_1} K(x,y)d\mu_h(y).$$

Based on the Martin integral representation, Naïm [**64**] extended the Fatou theorem. We say that $E \subset D$ is *minimally thin* at $y \in \Delta_1$ if $\widehat{R}^E_{K(\cdot,y)} \neq K(\cdot, y)$. We say that a function f on D has a *minimal fine limit* at y if the limit of f, except for a certain set minimally thin at y, exists. Minimal thinness may be considered to be a natural analogue of more familiar *thinness.* In general, a set $E \subset D$ is thin at $y \in D$ if and only if $\widehat{R}^E_{G(\cdot,y)} \neq G(\cdot, y)$. Irregularity of a domain with respect to the Dirichlet problem is equivalent to the thinness of the complement. The thinness is completely characterized by the the non-existence of a barrier and by the Wiener criterion.

Fatou-Naïm-Doob Theorem. *Let u and v be positive harmonic functions on D with Martin representation measures μ_u and μ_v, respectively. Then, for μ_v-a.e. $y \in \Delta_1$, the ratio u/v has minimal fine limit at y, which coincides with the Radon-Nikodým derivative $d\mu_h/d\mu_v(y)$.*

Armitage and Gardiner [**23**] is a good reference for the above accounts. The monograph covers almost all classical results in potential theory, from introductory materials to advanced subjects, such as minimal fine limits. Helms [**48**] is also a nice introduction to potential theory. Constantinescu and Cornea [**37**] is classical in connection with Riemann surfaces. Potential theory is closely related to probability theory. In this regard [**42**], [**70**], and [**26**] are nice references. A simple exposition of Naïm's theorem is included in the appendix of [**8**].

Now let us study the above general theorems for specific domains. If D is sufficiently smooth, then $G(x_0, y)$ is comparable to the distance function $\delta_D(y) = \operatorname{dist}(y, \partial D)$; the Martin kernel $K(x, y)$ is the inward normal derivative of the Green kernel times a positive function of y. Including this positive function in the representation measure, we may identify the Martin representation and the Poisson representation. Since $K(\cdot, y)$ vanishes continuously on $\partial D \setminus \{y\}$, it follows that $K(\cdot, y)$ is minimal, and the Martin boundary coincides with ∂D and consists of only minimal points. The arguments of Brelot and Doob [**34**] give a correspondence between minimal fine limits and non-tangential limits.

For a general domain, however, the Martin boundary may differ from the Euclidean boundary; non-minimal points may appear. No explicit criteria for minimal thinness are known. The more general domains we consider, the less precise properties we obtain. However, there are several interesting classes of domains between smooth domains and general domains, for which the Martin boundaries and minimal fine topologies can be described. In the study of such domains, we recognize that properties of harmonic functions in smooth domains, sometimes taken for granted, may heavily depend on the geometric nature of domains. Clarifying the relationship between the geometric structure of the domain and the properties of harmonic functions on it may help our intrinsic understanding of harmonic functions. How can we study such a relationship? The boundary Harnack principle stated in the next section will provide a powerful tool.

2. Boundary Harnack principle

Hereafter, A stands for a constant bigger than 1, whose value may change from one occurrence to the next. We say that two positive quantities f and g are comparable, and write $f \approx g$, if $A^{-1} f \leq g \leq Af$ with some $A > 1$. Let us recall the classical Harnack principle. Let K be a compact subset of a domain D. Then there exists $A > 1$, depending only on K and D, such that for all pairs of positive harmonic functions u and v on D,

$$\frac{u(x)/v(x)}{u(y)/v(y)} \leq A \quad \text{for } x, y \in K.$$

Let us, now, remove the assumption that K is a compact subset of D. What happens if K touches the boundary? No comparison of positive harmonic functions u and v will be available, because of the boundary values of u and v. To get around this, we impose the requirement $u = v = 0$ on the intersection of K and the boundary. More precisely, the boundary Harnack principle is formulated as follows.

Boundary Harnack Principle. *Let K be a compact set intersecting ∂D. Suppose $V \subset \mathbb{R}^n$ is an open set containing K. Then, there exists $A > 1$, depending only on K, V and D, such that*

$$\frac{u(x)/v(x)}{u(y)/v(y)} \leq A \quad \textit{for } x, y \in D \cap K,$$

whenever u and v are positive harmonic functions on D vanishing on $\partial D \cap V$.

The above comparison, referred to as the boundary Harnack principle (abbreviated BHP in the sequel), heavily depends on the geometric nature of the domain D. Let us give a remark on the condition that u vanishes on the boundary. We are going to study very general domains, including domains irregular with respect to the Dirichlet problem. For such domains we are not able to assume that u continuously vanishes on the boundary. The following somewhat weaker assumption is appropriate: u is bounded and u vanishes outside a polar set. This is written as $u = 0$ q.e. (quasi everywhere). Under this weak boundary condition, the same maximum principle holds as in the classical case (the Phragmén-Lindelöf principle). The Hausdorff dimension of a polar set is at most $n-2$, so that if a property holds q.e., then it holds a.e. with respect to the $n-2+\varepsilon$ dimensional Hausdorff measure for any $\varepsilon > 0$. The boundary condition of the Green function is understood to be $G(\cdot, y) = 0$ q.e. on ∂D. With this interpretation, the Green function always exists and the Martin boundary is defined, whenever the domain has a non-constant positive harmonic function.

3. Generalization of domains

Let us give a survey of generalizations of domains to which the Martin theory and the BHP are applicable. This survey may also offer some (incomplete) history of understanding of harmonicity without relying on explicit formulae.

3.1. Lipschitz domain. Suppose that a domain enjoys some smoothness assumption. For example, suppose that at each boundary point there exist two balls tangent at that point, one of which lies in the domain and the other outside the domain. Then, the comparison of the Green function for the domain and those for the ball and the complement of the closed ball gives that a positive harmonic function vanishing on a portion of the boundary is comparable to the distance function to the boundary near the portion. Thus the BHP readily follows in this case from this observation. However, problems arise for non-smooth domains. A domain whose boundary is represented locally as the graph of a Lipschitz continuous function is called a Lipschitz domain. The BHP was recognized for the first time in connection with Lipschitz domains.

Even for smooth domains, such as a ball and a half space, it is necessary to study Lipschitz domains, if we consider precise properties of harmonic functions. The following local Fatou theorem is a typical example. In Section 1 we stated the Fatou theorem, which can be strengthen to the following local form. Let us illustrate it for the upper half space $\mathbb{R}^n_+$ for simplicity.

Local Fatou Theorem. *Let h be a harmonic function on $\mathbb{R}^n_+$ non-tangentially bounded at $E \subset \partial\mathbb{R}^n_+$. Then, h has non-tangential limits at almost all points $x \in E$.*

Here, we say that h is non-tangentially bounded at E if for each $\xi \in E$, there is a truncated non-tangential cone $\Gamma(\xi)$ with vertex at ξ on which h is bounded. An elementary real analysis shows that the aperture and the size of $\Gamma(\xi)$ may be considered to be constant. Moreover, adding a suitable constant, we may assume that h is positive on $\Gamma(\xi)$. We observe that the union of non-tangential cones $\Gamma(\xi)$ is a Lipschitz domain. Thus the above local Fatou theorem reduces to the global Fatou theorem on a Lipschitz domain.

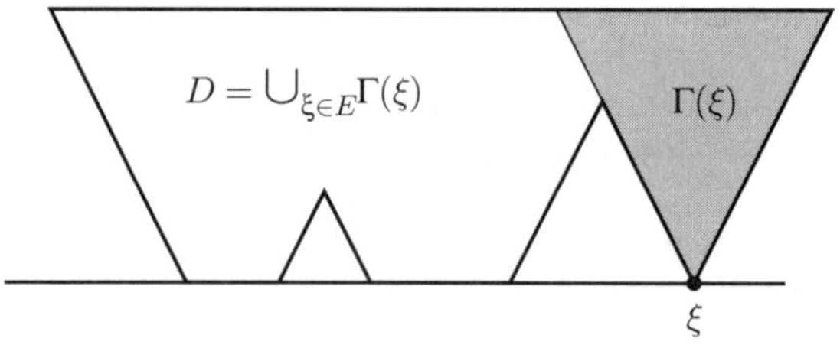

FIGURE 1. Union of non-tangential cones is a Lipschitz domain.

Carleson [**35**] proved this local Fatou theorem by showing the following Carleson estimate related to the BHP.

Carleson Estimate. *Let $\xi \in \partial D$, and let $r > 0$ be small. If u is a positive harmonic function in the union D of non-tangential cones, vanishing on $\partial D \cap B(\xi, Ar)$, then*

$$u(x) \le Au(\xi_r) \quad \text{for } x \in D \cap B(\xi, r),$$

where $\xi_r \in S(\xi, r) \cap D$ is a non-tangential point, i.e., $\delta_D(\xi_r) \approx r$.

Hunt and Wheeden [**50**], [**51**] applied the methods of Carleson to Lipschitz domains and argued, furthermore, to prove the Fatou theorem and to show that the Martin boundary coincides with the Euclidean boundary. In the late 1970s, the BHP for a Lipschitz domain was formulated explicitly. Ancona [**12**], Dahlberg [**41**] and Wu [**77**] proved the BHP almost at the same time by different methods. We distinguish two types of BHP's.

Scale Invariant Boundary Harnack Principle. *Let $\xi \in \partial D$, and let $r > 0$ be sufficiently small. If u and v are positive harmonic functions in $D \cap B(\xi, Ar)$ vanishing on $\partial D \cap B(\xi, Ar)$, then*

$$\frac{u(x)/v(x)}{u(y)/v(y)} \le A \quad \text{for } x, y \in D \cap B(\xi, r),$$

where $A > 1$ is independent of ξ, r, u and v. See Figure 6.

This scale invariant BHP, or uniform BHP, is important for applications. The BHP of Dahlberg was not scale invariant, though he employed methods of real analysis and proved some remarkable results: the harmonic measure is absolutely continuous with respect to the surface measure with density satisfying the Muckenhoupt A_∞ condition. His work played an important role in harmonic analysis on Lipschitz domains. On the other hand, Ancona studied not only harmonic functions but also positive solutions to general uniformly elliptic partial differential equations. His investigation proceeded from Lipschitz domains to more general non-smooth domains in Euclidean space, manifolds of negative curvature and discrete graphs ([**13**], [**14**], [**16**], [**17**], [**18**]). Wu studied harmonic functions very deeply. Her precise estimates of the harmonic measure of a box influenced the work of Jerison and Kenig and the box argument of Bass and Burdzy. See Section 5.

3.2. NTA domain. Jerison and Kenig [**52**] extended Lipschitz domains to NTA (Non-Tangentially Accessible) domains. They established the scale invariant BHP, the identification of the Martin boundary, the doubling property of the harmonic measure, the local Fatou theorem, and so on. Their methods are reminiscent of those for a Lipschitz domain. It was crucial for a Lipschitz domain that at each boundary point, there are interior and exterior cones of the same aperture and size with vertex at the boundary point. From the potential-theoretical point of view, the cones may be twisted, or even disconnected. Jerison and Kenig called a twisted cone a *corkscrew*. Let $0 < c < 1$. For $x, y \in D$, a *Harnack chain* from x to y in D is a sequence $\{B(x_j, c\delta_D(x_j))\}_{j=0}^N$ such that $x_0 = x$, $x_N = y$ and $B(x_{j-1}, c\delta_D(x_{j-1})) \cap B(x_j, c\delta_D(x_j)) \neq \emptyset$ for $j = 1, \dots, N$. The number N is referred to as the length of the Harnack chain. The value of c is not important and we may take $c = 1/2$. Jerison and Kenig say that a domain is NTA if there exist constants A and r_0 with the following properties:

(i) *Corkscrew condition.* For any $\xi \in \partial D$ and $0 < r < r_0$, there exists a point $x \in D \cap B(\xi, r)$ such that $\delta_D(x) \approx r$.
(ii) The complement of D satisfies the corkscrew condition.
(iii) *Harnack chain condition.* If $\varepsilon > 0$, and $x, y \in D$ with $\delta_D(x) > \varepsilon$, $\delta_D(y) > \varepsilon$ and $|x - y| < C\varepsilon$, then there exists a Harnack chain from x to y whose length depends on C, but not ε.

Under these assumptions, Jerison and Kenig succeeded in obtaining the BHP for an NTA domain in the same way as for a Lipschitz domain. As a result, the Martin boundary is homeomorphic to the Euclidean boundary. Nevertheless, NTA domains can be much more complicated than Lipschitz domains. There is an NTA domain whose boundary has Hausdorff dimension bigger than $n - 1$, so that the surface measure of the boundary may lose its meaning. Thus, Dahlberg's result does not extends to an NTA domain. Harmonic analysis on NTA domains is developed based on harmonic measure in place of surface measure.

3.3. Negatively curved manifold. Anderson and Schoen [**20**] proved that the Martin boundary associated with the Laplace-Beltrami equation of a simply connected complete Riemannian manifold is a sphere attached at infinity, provided the sectional curvature is bounded by a negative constant. Their proof relied on a special superharmonic function and complicated calculations. Ancona [**16**] gave an alternative proof based on techniques in potential theory stated as above. His method is easier and more applicable. Let $\mathcal{L}$ be a uniformly elliptic operator on a manifold. Suppose $\mathcal{L}$ is *weakly coercive*, i.e., there is a positive superharmonic function with respect to $\mathcal{L} + \varepsilon I$ for some $\varepsilon > 0$. Under this assumption, Ancona showed the BHP and identified the Martin boundary. A domain in Euclidean space is weakly coercive if and only if it has a *strong barrier* s, i.e., not only is s positive and superharmonic but also it is more superharmonic, in the sense that

$$\Delta s(x) + \frac{\varepsilon}{\delta_D(x)^2} s(x) \leq 0 \quad \text{for some } \varepsilon > 0.$$

The existence of a strong barrier is equivalent to *Hardy's inequality*:

$$\int_D \left| \frac{\varphi(x)}{\delta_D(x)} \right|^2 dx \leq c_D \int_D |\nabla \varphi(x)|^2 dx \quad \text{for } \varphi \in C_0^\infty(D),$$

which, in turn, holds if the boundary enjoys the *Capacity Density Condition* (CDC):

$$\frac{\mathrm{Cap}_{B(x,2r)}(\partial D\cap B(x,r))}{\mathrm{Cap}_{B(x,2r)}(B(x,r))}\geq c>0\quad\text{for } x\in\partial D \text{ and } 0<r<r_0.$$

Here, $\mathrm{Cap}_{B(x,2r)}(E)$ is the Green capacity of E with respect to $B(x,2r)$ ([**15**]). Recall that the famous Wiener criterion for the regularity of a boundary point with respect to the Dirichlet problem is

$$\sum_{j=0}^{\infty}\frac{\mathrm{Cap}_{B(x,2^{1-j})}(D^c\cap B(x,2^{-j}))}{\mathrm{Cap}_{B(x,2^{1-j})}(B(x,2^{-j}))}=\infty.$$

We easily observe that $\mathrm{Cap}_{B(x,2^{1-j})}(D^c\cap B(x,2^{-j}))$ can be replaced by

$$\mathrm{Cap}_{B(x,2^{1-j})}(\partial D\cap B(x,2^{-j})).$$

Hence, if the boundary satisfies the CDC, then the domain is regular and has a uniform barrier.

For a planar domain, the CDC of the boundary, the existence of a strong barrier, the Hardy inequality, and the uniform perfectness of the boundary are equivalent to each other. In particular, if a domain has a strong barrier, then it is regular. For a domain in a higher dimensional space, the CDC of the boundary implies the existence of a strong barrier, but the converse does not hold. A punctured ball is irregular and does not satisfy the CDC and yet has a strong barrier ([**15**]). Characterization of domains having a strong barrier is open in the higher dimensional case.

Ancona's method is widely applicable. Ancona himself applied it to the Martin boundary of a discrete graph ([**17**]). Arai [**21**], [**22**] extended it to degenerate elliptic equations. Bonk, Heinonen, and Koskela [**33**] studied it in connection with Gromov hyperbolicity, as we shall see in the next section.

One may wonder whether conditions such as coercivity and existence of a strong barrier are essential for the identification of the Martin boundary and the BHP. In fact, it turns out that there is a certain family of domains, without such conditions, for which the BHP holds.

3.4. Irregular domains. Bass and Burdzy [**27**] rather recently established the BHP for a Hölder domain[1] and a John domain by using probabilistic techniques. In view of the "Lebesgue spine", a Hölder domain need not be regular with respect to the Dirichlet problem for the higher dimensional case. The heart of Bass and Burdzy's argument requires no exterior conditions; only interior conditions yield the BHP. From this point of view the Bass-Burdzy method is remarkable. Unfortunately, their BHP is not scale invariant and is not sufficient for the identification of the Martin boundary. We now raise a question: Is there a domain more general than an NTA domain for which the scale invariant BHP holds and whose Martin boundary coincides with the Euclidean one? In the succeeding sections, we shall, indeed, prove that a uniform domain is required. In the procedure, we shall understand the Bass-Burdzy probabilistic argument in a purely analytic way. The

[1]Here, a Hölder domain is a domain whose boundary is locally given by the graph of a Hölder continuous function. Smith and Stegenga [**73**] said that a domain is Hölder if it satisfies the quasi-hyperbolic boundary condition. A John domain is a Hölder domain in the sense of Smith and Stegenga. Very complicatedly, several related papers use this notation ([**74**], [**24**], [**47**]).

crux of their argument can be interpreted as the estimate of the Green function by a decomposition of a domain.

4. Uniform domain and John domain

A uniform domain is a domain closely related to quasiconformal mappings. It has an intimate connection with the extendability of BMO functions and Sobolev functions as well. See [**46**], [**53**], [**54**], [**45**], [**75**], [**69**], and so on. There are several equivalent definitions for a uniform domain. Here we adopt the following: We say that D is a *uniform domain* if every pair of points $x, y \in D$ can be connected by a curve $\gamma \subset D$ such that

$$\ell(\gamma) \le A|x-y|,$$
$$\min\{\ell(\gamma(x,z)), \ell(\gamma(z,y))\} \le A\delta_D(z) \quad \text{for } z \in \gamma,$$

where $\gamma(x,z)$ is the subarc of γ connecting x and z. This can be visualized as the existence of an object, cigar-like, plumping in the middle with endpoints at x and y whose length is bounded by $|x-y|$ up to a multiplicative constant in the domain. Thus, this condition is referred to as a *cigar condition.* A Lipschitz domain and an NTA domain are uniform domains. Roughly speaking, a uniform domain is a domain satisfying only the interior conditions for an NTA domain.

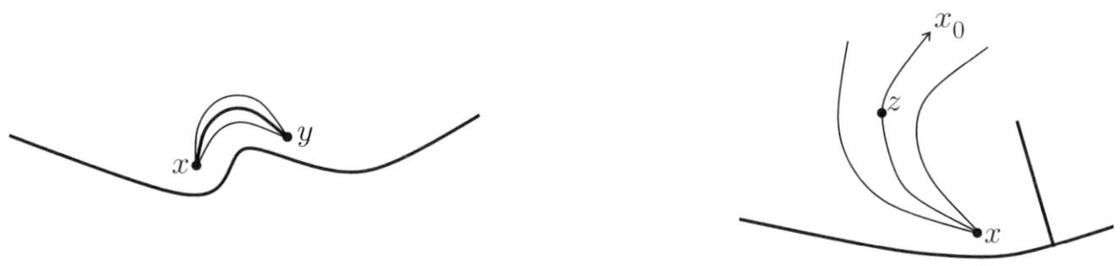

FIGURE 2. Uniform domain (cigar condition) and John domain (carrot condition).

While a pair of points x and y can move freely in D for the definition of a uniform domain, the same condition with one point fixed, say $y = x_0$, and one point moving, $x \in D$, gives a *John domain.* More precisely, we say that D is a John domain with John center at x_0 and John constant c_J if every point $x \in D$ can be connected to x_0 by a curve $\gamma \subset D$ such that

$$\delta_D(z) \ge c_J \ell(\gamma(x,z)) \quad \text{for } z \in \gamma.$$

We may replace the center x_0 by a fixed compact set of D. In general, the John constant c_J is less than 1. If c_J is getting close to 1, we may consider that D is getting smooth. The definition of a John domain implies that for every $x \in D$, there is a twisted cone with vertex at x and fixed aperture toward x_0. Because of its geometric nature, the condition is referred to as a *carrot condition.* It is known that a cigar condition without bounded turning condition, $\ell(\gamma) \le A|x-y|$, is equivalent to the carrot condition.

Balogh and Volberg [**25**] introduced a *uniformly John domain,* a domain intermediate between a John domain and a uniform domain. This is given by relaxing the bounded turning condition $\ell(\gamma) \le A|x-y|$ for a uniform domain to $\ell(\gamma) \le A\rho_D(x,y)$, where $\rho_D(x,y)$ is the internal metric, or the inner diameter metric, between x and y defined by the infimum of the diameters of curves connecting

x and y in D. The inner length distance $\lambda_D(x,y)$ between x and y is defined by the infimum of the lengths of curves connecting x and y in D. Bonk, Heinonen and Koskela [**33**] say that D is an *inner uniform domain* if every pair of points $x, y \in D$ can be connected by a curve $\gamma \subset D$ such that

$$\begin{aligned} &\ell(\gamma) \le \lambda_D(x,y), \\ &\min\{\ell(\gamma(x,z)), \ell(\gamma(z,y))\} \le A\delta_D(z) \quad \text{for } z \in \gamma, \end{aligned}$$

connecting x and y. Obviously, $|x-y| \le \rho_D(x,y) \le \lambda_D(x,y)$, and hence a uniformly John domain is an inner uniform domain. It turns out, however, that the converse holds, since $\lambda_D(x,y) \le A\rho_D(x,y)$ for a John domain (Väisälä [**76**, Theorem 3.4]).

In terms of the *quasihyperbolic metric*

$$k_D(x,y) = \inf_{\widetilde{xy}} \int_{\widetilde{xy}} \frac{ds}{\delta_D(z)}$$

with the infimum being taken over all curves $\widetilde{xy} \subset D$ connecting x and y, a uniform domain is characterized by

$$k_D(x,y) \le A \log\left(\frac{|x-y|}{\min\{\delta_D(x), \delta_D(y)\}} + 1\right)$$

(Gehring and Osgood [**46**]). We see that a John domain satisfies a *quasihyperbolic boundary condition*:

$$k_D(x, x_0) \le A \log \frac{\delta_D(x_0)}{\delta_D(x)} + A.$$

It is known that a bounded simply connected planar domain is mapped to the unit disc by a Hölder continuous mapping if and only if the domain satisfies the quasihyperbolic boundary condition ([**31**]). From this fact, Smith and Stegenga [**73**] called a domain satisfying the quasihyperbolic boundary condition a Hölder domain. The following summarizes the definitions so far:

$$\begin{aligned} \text{Smooth} &\subsetneq \text{Lipschitz} \subsetneq \text{NTA} \\ &\subsetneq \text{Uniform} \subsetneq \text{Uniformly John} = \text{Inner uniform} \subsetneq \text{John} \\ &\subsetneq \text{Quasihyperbolic boundary condition holds.} \end{aligned}$$

Whereas NTA and smoother domains satisfy exterior conditions and hence they are regular, uniform domains and more general domains satisfy only interior conditions and may be irregular. Many known results used exterior conditions. For example, [**25**] proved the BHP for a planar uniformly John domain under the additional assumption that it has a uniformly perfect boundary. Bonk, Heinonen, and Koskela [**33**] proved that an inner uniform domain has Gromov hyperbolicity; and under the existence of a strong barrier, they identified the Martin boundary following the scheme of [**16**]. As stated in the previous section, an analytic interpretation of [**27**] enables us to prove the following theorem without any additional exterior condition ([**5**]).

THEOREM 1. *A bounded uniform domain has the scale invariant BHP. Moreover, its Martin boundary coincides with the Euclidean boundary and consists only of minimal points.*

The following is an immediate corollary.

COROLLARY 2. *Let D be a bounded uniform domain and V an open set intersecting ∂D. If u and v are positive harmonic functions on D, bounded on $D \cap V$ and vanishing q.e. on $\partial D \cap V$, then the ratio u/v extends to $\overline{D} \cap V$ as a Hölder continuous function.*

Since D may be irregular in general, u and v may be discontinuous at the boundary. Nevertheless, the ratio u/v is Hölder continuous. The corollary can be proved with the aid of the BHP in the same way as de Giorgi and Moser ([**58**]) proved the (interior) Hölder continuity from the Harnack inequality. Moreover, if the Euclidean metric is replaced by the internal metric, then results similar to Theorem 1 and Corollary 2 hold for a uniformly John domain ([**10**]).

5. Box argument

Let us touch the heart of the proof of the BHP for a uniform domain. A Lipschitz domain satisfies the exterior cone condition, so that it has a uniform barrier. This is one of the main ingredients of the proof of Carleson and succeeding arguments. On the other hand, a uniform domain and a John domain satisfy only interior conditions, and they need not have a uniform barrier and may be even irregular. Moreover, a surface ball $\partial D \cap B(\xi, r)$ may be polar and the harmonic measure need not be doubling. These are crucial differences between an NTA domain and a uniform domain.

For a John domain, it is important that the carrot condition gives some information on the *width* of the domain near the boundary. Observe that each point $x \in U_r = \{x \in D : \delta_D(x) < r\}$ can be connected to x_0 by a curve with the carrot condition. Hence, $B(x, Ar) \setminus U_r$ includes a ball of radius r provided A is large enough. Let us consider the harmonic measure $\omega = \omega(\cdot, S(x, Ar) \cap U_r, B(x, Ar) \cap U_r)$. Note that ω is a bounded harmonic function on $B(x, Ar) \cap U_r$ such that $\omega = 1$ q.e. on $S(x, Ar) \cap U_r$ and $\omega = 0$ q.e. on $B(x, Ar) \cap \partial U_r$. We extend ω to $B(x, Ar) \setminus U_r$ by $\omega = 0$. Then the extended function is subharmonic on $B(x, Ar)$. Hence the volume mean inequality implies that. $\omega \leq 1 - \varepsilon$ on $B(x, 2r) \cap U_r$, with $\varepsilon > 0$ being independent of r. Applying the maximum principle and this inequality repeatedly, we obtain the following.

LEMMA 3. *There are positive constants A_1 and A_2, depending only on the domain, such that if $0 < Ar < R$, then*

$$\omega(y, S(x, R) \cap U_r, B(x, R) \cap U_r) \leq \exp(A_1 - A_2 \frac{R}{r}) \quad \textit{for } y \in B(x, 2r) \cap U_r.$$

For the above lemma, $B(x, Ar) \setminus U_r$ need not include a ball of radius r. Instead,

$$\frac{\mathrm{Cap}_{B(x,2Ar)}(B(x, Ar) \setminus U_r)}{\mathrm{Cap}_{B(x,2Ar)}(B(x, Ar))} \geq c > 0 \quad \text{for } x \in U_r$$

is sufficient. In this case, we say that the *capacitary width* of U_r is less than Ar. Lemma 3 implies that the harmonic measure decreases rapidly when the capacitary width decreases. The same estimate hold for a domain, not necessarily John, but satisfying the CDC. This is the common point to our method and the classical method going back to Carleson. The crux of Theorem 1 is the comparison between the Green function and the harmonic measure based on Lemma 3.

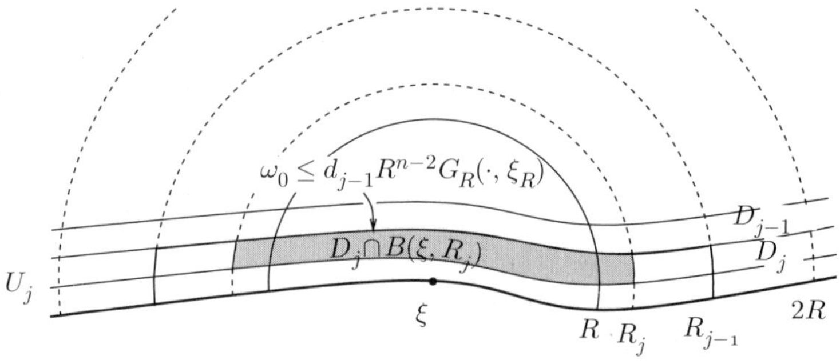

FIGURE 3. Maximum principle over $U_j \cap B(\xi, R_{j-1})$.

LEMMA 4. *Let $\xi \in \partial D$, and let $R > 0$ be sufficiently small. Take a nontangential point $\xi_R \in D \cap S(\xi, 4R)$, i.e., $\delta_D(\xi) \approx R$. Let $A > 10$ be sufficiently large and let G_R be the Green function for $D \cap B(\xi, AR)$. Then*

$$\omega(y, D \cap S(\xi, 2R), D \cap B(\xi, 2R)) \le AR^{n-2}G_R(y, \xi_R) \quad \textit{for } y \in D \cap B(\xi, R).$$

PROOF. The length of the Harnack chain along a curve joining x and x_R with the cigar condition is bounded by $A \log R/\delta_D(x)$. Harnack's inequality gives a $\lambda > 0$ such that

$$R^{n-2}G_R(x, \xi_R) \ge A\left(\frac{\delta_D(x)}{R}\right)^\lambda.$$

For $j \ge 0$ we put $D_j = \{x \in D : \exp(-2^{j+1}) \le R^{n-2}G_R(x, \xi_R) < \exp(-2^j)\}$ and $U_j = \{x \in D : R^{n-2}G_R(x, \xi_R) < \exp(-2^j)\}$. Then $U_j \subset \{x \in D : \delta_D(x) < AR\exp(-2^j/\lambda)\}$. Take $R_j \downarrow R$ decreasing so slowly that $R_{j-1} - R_j \approx j^{-2}R$. For simplicity let $\omega_0 = \omega(\cdot, D \cap S(\xi, 2R), D \cap B(\xi, 2R))$ and

$$d_j = \sup_{x \in D_j \cap B(\xi, R_j)} \frac{\omega_0(x)}{R^{n-2}G_R(x, \xi_R)}.$$

Then the lemma follows from $\sup_{j \ge 0} d_j \le A < \infty$.

Apply the maximum principle over $U_j \cap B(\xi, R_{j-1})$ to obtain

$$\omega_0 \le \omega(\cdot, U_j \cap S(\xi, R_{j-1}), U_j \cap B(\xi, R_{j-1})) + d_{j-1}R^{n-2}G_R(\cdot, \xi_R)$$

on $U_j \cap B(\xi, R_{j-1})$. Divide both sides by $R^{n-2}G_R(\cdot, \xi_R)$ and take the supremum over $D_j \cap B(\xi, R_j)$. Then Lemma 3 implies

$$d_j \le A\exp\left(2^{j+1} - Aj^{-2}\exp\left(\frac{2^j}{\lambda}\right)\right) + d_{j-1}.$$

Since $\sum_j \exp\left(2^{j+1} - Aj^{-2}\exp(2^j/\lambda)\right)$ is convergent, we have $\sup_{j \ge 0} d_j \le A < \infty$. The lemma is proved. □

PROOF OF THEOREM 1. Since u is a positive bounded harmonic function on $D \cap B(\xi, AR)$ vanishing q.e. on $\partial D \cap B(\xi, AR)$, we obtain a potential representation:

$$u(x) = \int_{D \cap S(\xi, 6R)} G_R(x, y)d\mu(y) \quad \text{for } x \in D \cap B(\xi, 6R).$$

Take $x \in D \cap B(\xi, R)$ and $y \in D \cap S(\xi, 6R)$. Moreover, let $x^* \in D \cap S(\xi, R)$ and $y^* \in D \cap S(\xi, 6R)$ be nontangential points. Observe that $G_R(x, y)$ with either x

or y fixed is a harmonic function of the other argument for $x \neq y$ vanishing on the boundary. Multiply the trivial inequality $G_R(x,y) \le AR^{2-n}$ by the harmonic measure and apply the maximum principle. Then we have from Lemma 4

$$G_R(x,y) \le AR^{2-n}R^{n-2}G_R(x,y^*)R^{n-2}G_R(x^*,y) \le A\frac{G_R(x,y^*)G_R(x^*,y)}{G_R(x^*,y^*)},$$

where $G_R(x^*,y^*) \approx R^{2-n}$ is used. The reverse inequality is derived rather easily. Integration with respect to μ yields

$$u(x) \approx \frac{G_R(x,y^*)}{G_R(x^*,y^*)}\int_{D\cap S(\xi,6R)} G_R(x^*,y)d\mu(y) = \frac{G_R(x,y^*)}{G_R(x^*,y^*)}u(x^*).$$

Since v has a similar estimate, the theorem follows. □

REMARK 5. The series at the end of the proof of Lemma 4 converges with a good margin. This may suggest an extension of the lemma to more general domains. In fact, by a probabilistic method, Bass and Burdzy [**29**] identified the Martin boundary of a Hölder domain with modulus of continuity slightly worse than Lipschitz. Its analytic counterpart and the BHP remain open.

6. Fatou theorem

Let us return to our starting point, the Fatou theorem. Jerison and Kenig [**52**] proved the local Fatou theorem for an NTA domain. We can show that a uniform domain is the most general domain for which the Fatou theorem holds ([**5**]).

THEOREM 6. (i) *A bounded uniform domain D satisfies the global Fatou theorem, i.e., a positive harmonic function u on D has nontangential limit at almost every boundary point with respect to the harmonic measure.*

(ii) *There is a bounded uniform domain D failing to satisfy the local Fatou theorem, i.e., there is a harmonic function u nontangentially bounded at almost every boundary point and yet having no nontangential limit at almost every boundary point with respect to the harmonic measure.*

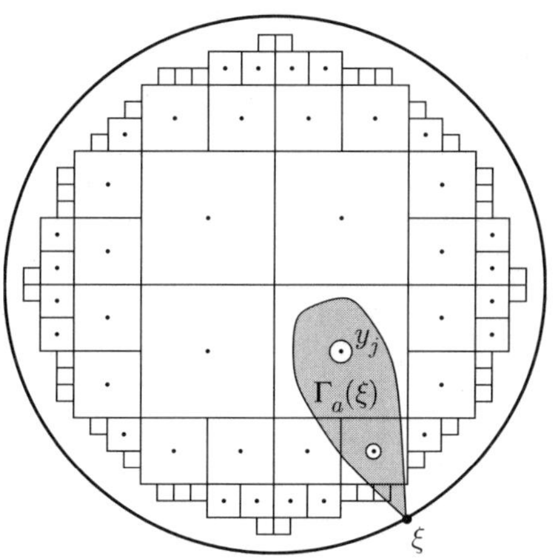

FIGURE 4. Uniform domain $D = B(0,1) \setminus \bigcup_{j=1}^{\infty}\{y_j\}$ and nontangential cone $\Gamma_\alpha(\xi)$.

The first assertion follows from the classical Fatou-Naïm-Doob theory concerning the boundary behavior in the framework of Martin spaces, since u is positive on D. On the other hand, for the last assertion, we consider the Whitney decomposition $\bigcup_{j=1}^{\infty} Q_j$ of the unit ball $B(0,1)$ with center y_j of Q_j. Let $D = B(0,1)\setminus\bigcup_{j=1}^{\infty}\{y_j\}$. Then D is a uniform domain, and a nontangential cone $\Gamma_\alpha(\xi) = \{x \in D : |x-\xi| < (1+\alpha)\delta_D(x)\}$ with vertex at $\xi \in S(0,1)$ may be depicted as in Figure 4. We can construct a harmonic function u having a singularity of order $-j$ at y_j such that for each $\xi \in S(0,1)$, if the aperture α is small, then u is bounded on the nontangential cone $\Gamma_\alpha(\xi)$; if the aperture α is large, then the nontangential cone $\Gamma_\alpha(\xi)$ has a sequence to ξ along which u diverges. This strange phenomenon seems to be caused by lack of the doubling property of the harmonic measure on the boundary of the domain.

7. Martin boundary points

So far we have generalized domains and arrived at a uniform domain, whose Martin boundary coincides with the Euclidean boundary. There are somewhat different studies, which investigate not the whole Martin boundary but the Martin boundary over a fixed point on the Euclidean boundary. A typical example is the Picard principle dealing with the Martin boundary over the origin associated with the Schrödinger equation: $-\Delta u + Pu = 0$ on a punctured unit disk. Nakai, Tada, and others ([**65**], [**66**], [**67**], [**68**] and references therein) give very precise estimates of the number of Martin boundary points over the origin. Murata [**59**], [**60**], [**61**], [**62**], [**63**] has made intensive and extensive studies of the Martin boundary at infinity associated with the Schrödinger equation for the whole space, skew products and manifolds, and so on. Murata's works are not confined to elliptic equations, but are related to parabolic equations. Unfortunately, they are beyond the scope of this survey. We shall restrict ourselves to domains in Euclidean space and discuss some more results.

7.1. Denjoy domain. Let E be a closed set on the hyperplane $\{x = (x_1, \dots, x_n) : x_n = 0\}$. The complement $\mathbb{R}^n \setminus E$ is referred to as a *Denjoy domain.* If E is the hyperplane itself, then the complement is not a domain but consists of two half spaces. If we remove a bounded set from the hyperplane, then the complement is a Denjoy domain having two linearly independent positive harmonic functions vanishing on finite boundary points. In the terminology of the Martin boundary, there are two minimal Martin boundary points at infinity. Our intuition suggests that this is the case if E is large enough at infinity. In fact, Benedicks [**32**] proved the following celebrated theorem.

THEOREM 7. *A Denjoy domain has either one or two minimal Martin boundary points at infinity. Let K_x be the cube with center at x and sides of length $c|x|$, parallel to the coordinate axes, where $0 < c < 1$. Let $\beta_E(x) = \omega(x, \partial K_x, K_x \setminus \partial D)$. Then the Denjoy domain has two minimal Martin boundary points at infinity if and only if*

$$\int_{|x|\geq 1} \frac{\beta_E(x)}{|x|^{n-1}} dx_1 \cdots dx_{n-1} < \infty.$$

This theorem and the monotonicity of the harmonic measure β_E give the next corollary.

COROLLARY 8 (Monotonicity). *Let $E \subset F$ be closed sets on the hyperplane. If the Denjoy domain $\mathbb{R}^n \setminus E$ has two minimal Martin boundary points at infinity, then so does the Denjoy domain $\mathbb{R}^n \setminus F$.*

If there are two minimal Martin boundary points at infinity, then there exist non-minimal Martin boundary points by the connectedness of the Martin boundary. Benedicks' criterion gives a necessary and sufficient condition in terms of the harmonic measure, whose estimate is not obvious. Segawa [**71**], [**72**] and Gardiner [**44**] give more explicit conditions in terms of the Lebesgue measure.

The Kelvin transform converts the Benedicks theorem to a criterion for the number of minimal Martin boundary points over the origin of a bounded domain. Let E be a compact set on a Lipschitz surface Σ and suppose B is an open ball containing E. We call $B \setminus E$ a Lipschitz Denjoy domain. If Σ is smooth, then an analogue of Benedicks' theorem holds (Chevallier [**36**]). The first assertion of the Benedicks theorem is valid for a general Lipschitz Denjoy domain, i.e., the number of minimal Martin boundary points over a Euclidean boundary point is at most two (Ancona [**14**] and Chevallier [**36**]). However, the second assertion no longer holds. One might think, intuitively, that the larger the boundary, the more minimal Martin boundary points. This is not the case. Ancona [**18**] constructed a counterexample to Corollary 8 for a Lipschitz Denjoy domain, i.e., there are closed sets $E \subset F \subset \Sigma$ such that $B \setminus F$ has one minimal Martin boundary point over the origin, whereas $B \setminus E$ has two. Thus, there is no Benedicks type criterion for a Lipschitz Denjoy domain.

The following *weak boundary Harnack principle* is used for the proof of the fact that the number of minimal Martin boundary points is one or two: Let Σ be a Lipschitz surface containing the origin. For small $r > 0$ there are points x_r^+ and x_r^- on $S(0,r)$ whose distances to Σ are comparable to r and Σ lies between x_r^+ and x_r^-. Suppose h_0, h_1 and h_2 are positive harmonic functions on $D = B \setminus E$ such that $h_0 = h_1 = h_2 = 0$ q.e. on ∂D and bounded apart from the origin. Then

$$h_0(x) \le A\left(\frac{h_0(x_r^+)}{h_1(x_r^+)}h_1(x) + \frac{h_0(x_r^-)}{h_2(x_r^-)}h_2(x)\right) \quad \text{for } x \in D \setminus B(0,r).$$

Let $\xi \in \partial D$. We say that a positive harmonic function h on D is a *kernel function at* ξ if h vanishes q.e. on ∂D, is bounded apart from ξ, and $h(x_0) = 1$. Here $x_0 \in D$ is a reference point. The weak BHP asserts that among three kernel functions at the origin, there is one kernel function bounded by the sum of the others up to a multiplicative constant. This observation readily implies that the number of minimal Martin boundary points over the origin is at most two.

7.2. Sectorial domain and John domain. Cranston and Salisbury [**40**] called a planar domain whose boundary lies on finitely many rays leaving the origin a *sectorial domain.* See Figure 5. They proved a result corresponding to Benedicks' theorem and related probabilistic results for a sectorial domain. Lömker [**56**] called a higher dimensional analogue a *quasisectorial domain* and gives similar results with respect to the Schrödinger equation. Since a Denjoy domain, a sectorial domain and a quasisectorial domain are all John domains, it may be natural to study the minimal Martin boundary points of a John domain. As Ancona's counterexample illustrates, precise results similar to the last assertion of the Benedicks theorem need not hold without strict assumptions on the location of the boundary. Nonetheless we can prove the following result [**9**] by establishing a weak BHP.

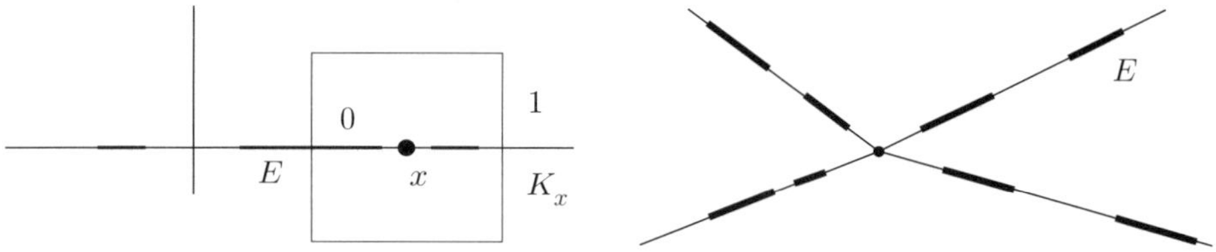

FIGURE 5. Denjoy domain (harmonic measure β_E) and sectorial domain.

THEOREM 9. *The number of minimal Martin boundary points over a Euclidean boundary point of a bounded John domain is finite and estimated by the John constant c_J. Moreover, if the John constant c_J is bigger than $\sqrt{3}/2$, then the number of minimal Martin boundary points over a boundary point is at most two.*

The theorem sounds reasonable from the point of view that the smoother the domain is, the closer to 1 the John constant c_J is. However, the most interesting phenomenon, that there is one Martin boundary point at each Euclidean boundary point and the Martin boundary coincides with the Euclidean boundary, cannot be derived from the theorem. In this direction, Ancona [**13**] showed that the union D of a family of open balls with the same radius ρ_0 has one minimal Martin boundary point at a Euclidean boundary point ξ if whenever D includes two open balls B_1 and B_2 with radius ρ_0 tangential to each other at ξ, D includes a truncated circular cone $\Gamma_\theta(\xi, y) \cap B(\xi, r)$ for some $\theta > 0$, $r > 0$ and y in the hyperplane tangent to B_i at ξ. Ancona's argument is based on subtle estimates of the Green function, for which it is crucial that D is the union of balls with the same radius. Applying the study on the minimal Martin boundary points of a John domain, we can prove that the union of non-shrinking convex sets has one minimal Martin boundary point at each Euclidean boundary point and that the Martin boundary coincides with the Euclidean boundary, provided D satisfies an appropriate condition corresponding to Ancona's cone condition ([**9**]).

8. Applications and related topics

8.1. Minimally thin sets. Let us give an explicit criterion for minimally thin sets, similar to the Wiener criterion for thin sets, in order to understand the Fatou-Naïm-Doob theorem in a more concrete fashion. The usual Wiener criterion is given as the convergence of the series of capacities of the intersections of concentric spherical shells and the set under consideration. A similar Wiener type criterion can be given for minimally thin sets. In place of the logarithmic capacity and the Newtonian capacity, the Green capacity, defined in terms of the Green function, is involved in the series as long as the same intersection is taken. Since the Green function is implicit, so is the Green capacity. More explicitly, quantities are wanted to describe minimal thinness. This is possible. Taking the intersection with Whitney cubes instead of concentric spherical shells, we can describe the Wiener type criterion for minimal thinness in terms of the logarithmic capacity and the Newtonian capacity. In general, capacities are subadditive, i.e., the capacity of a set is dominated by the summation of the capacities of the partitions of the set. For the Green capacity and the Whitney decomposition, an opposite inequality up to a multiplicative constant holds under a certain condition on the domain (quasiadditivity). In each Whitney cube, the Green capacity can be estimated by the logarithmic capacity and the Newtonian capacity; and hence a Wiener type criterion for minimally thin sets can be stated in terms of the logarithmic capacity and the Newtonian capacity. For simplicity we shall restrict ourselves to a smooth domain D in the Euclidean space of dimension greater than or equal to three. Let $\{Q_j\}$ be the Whitney decomposition of D. Then $E \subset D$ is minimally thin at $x \in \partial D$ if and only if

$$\sum \frac{\operatorname{dist}(Q_j, \partial D)^2}{\operatorname{dist}(Q_j, x)^n} \operatorname{Cap}(E \cap Q_j) < \infty,$$

where Cap is the Newtonian capacity. The proof of this claim involves the BHP and the Hardy inequality ([**1**] and [**8**, Corollary 4.4]). A similar but more complicated condition can be derived for an NTA domain. Not only the identification of the Martin boundary, but also the investigation of the properties of minimally thin sets are necessary for the complete understanding of positive (super)harmonic functions on the domain. It is an open problem to study minimally thin sets for manifolds and domains more general than an NTA domain.

8.2. 3G inequality. From the BHP, Cranston, Fabes, and Zhao [**38**] derived the following 3G inequality of the ratio of three Green functions for a Lipschitz domain in a Euclidean space of dimension greater than or equal to three:

$$\frac{G(x,y)G(y,z)}{G(x,z)} \le A(|x-y|^{2-n} + |y-z|^{2-n}) \quad \text{for } x, y, z \in D.$$

They applied the inequality to the probabilistic analysis for the Schrödinger equation. Observe that the denominator $G(x, z)$ tends to 0 as x and z approach the boundary, and so does the numerator $G(x, y)G(y, z)$. The BHP asserts that they cancel, and hence the ratio is bounded. We can prove the 3G inequality for a uniformly John domain. Since the Martin boundary of a uniformly John domain differs from the Euclidean boundary, there is no relationship between the 3G inequality and the Martin boundary. Surprisingly, Bass and Burdzy [**30**] proved the 3G inequality for arbitrary bounded planar domains by probabilistic methods. The analytic proof and the 3G inequality for general domains in the higher dimensional case remain open.

8.3. Integrability of harmonic functions. In view of the Herglotz theorem we see that a positive harmonic function on the unit ball is L^p-integrable with $p < n/(n-1)$ and the bound $p < n/(n-1)$ is sharp. How about a general domain? Let us consider, for example, the first quadrant $\{z = x + iy : x > 0, y > 0\}$ of the plane. An easy calculation shows that $h(z) = xy|z|^{-4}$ is positive and harmonic there. Since h is homogeneous of degree -2, it is not L^1-integrable near the origin. Thus, the bound $p < n/(n-1)$ must vary for a domain with edges. Let us call a domain whose boundary is locally represented by the graph of a k-Lipschitz function on $\mathbb{R}^{n-1}$, a k-Lipschitz domain. We shall see that the BHP yields the following theorem ([**2**]).

THEOREM 10. *Let D be a bounded k-Lipschitz domain in $\mathbb{R}^n$. Then there exists a constant p_k, depending only on k and the dimension, such that every positive harmonic function on D is L^p-integrable with $p < p_k$. Moreover, $p_k \to n/(n-1)$ as $k \to 0$ and $p_k = 1$ if $k = 1/\sqrt{n-1}$.*

PROOF. Fix $x_0 \in D$ and let $g(x) = G(x, x_0)$ and $D_t = \{x \in D : g(x) > t\}$. Observe that $g(x) - t$ is the Green function for D_t with pole at x_0. For $b > 0$ and $\varepsilon > 0$, the coarea formula yields

$$\int_{\{g(x)<b\}} h(x)g(x)^{\varepsilon-1}|\nabla g(x)|^2 dx = \int_0^b t^{\varepsilon-1}dt \int_{\partial D_t} h|\nabla g| d\sigma = c_n \frac{b^\varepsilon}{\varepsilon} h(x_0),$$

which gives a weighted integrability of h. This holds for a general regular domain.

Let us now investigate the weight $g(x)^{\varepsilon-1}|\nabla g(x)|^2$ with the aid of the geometric nature of the domain. The Harnack inequality readily yields $|\nabla g(x)| \le Ag(x)/\delta_D(x)$. We need an inequality that is reverse to this one in a certain sense. To this end, let $\{Q_j\}$ be the Whitney decomposition of D and let y_j be the center of Q_j. Take $y_j^* \in \partial D$ such that $\delta_D(y_j) = |y_j - y_j^*|$. Then the BHP yields

$$g \le Ag(y_j)\omega(\cdot, D \cap S(y_j^*, \delta_D(y_j)), D \cap B(y_j^*, \delta_D(y_j)) \quad \text{on } D \cap B(y_j^*, A^{-1}\delta_D(y_j)).$$

The harmonic measure is estimated by the exterior cone condition:

$$\omega(y, D \cap S(y_j^*, \delta_D(y_j)), D \cap B(y_j^*, \delta_D(y_j)) \le A \left(\frac{|y - y_j^*|}{\delta_D(y_j)} \right)^\beta,$$

for some $\beta > 0$. We find a point y_j' on the line segment $\overline{y_j y_j^*}$ such that $g(y_j') \le \frac{1}{2} g(y_j)$ and $\delta_D(y_j') \ge A^{-1}\delta_D(y_j)$. Since g is positive and harmonic on $B(y_j, \delta_D(y_j)) \subset D$, we find a point $y_j'' \in Q_j$ such that $g(y_j'') \le (1 - \eta)g(y_j)$. In fact, if otherwise, the dilation and translation would give constants $0 < c_1 < c_2 < 1$ and a sequence of positive harmonic functions g_j on the unit ball such that $g_j(0) = 1$, $\inf_{B(0,c_2)} g_j \le 1/2$ and $\lim_{j\to\infty}(\inf_{B(0,c_1)} g_j) = 1$. Taking a subsequence if necessary, we may assume that g_j converges to a positive harmonic function g_0, which would satisfy $\inf_{B(0,c_2)} g_0 \le 1/2$ and $g_0(0) = 1 = \inf_{B(0,c_1)} g$. The last equality, the minimum principle and the analyticity of harmonic functions would yield $g_0 \equiv 1$. This would contradict the first inequality.

Now letting $\tau > 0$ be sufficiently small, we obtain

$$|\{x \in Q_j : |\nabla g(x)| \ge \tau g(y_j)/\delta_D(y_j)\}| \ge A^{-1}|Q_j|.$$

On the other hand, there is a positive constant α, depending on the Lipschitz constant k, such that $g(x) \ge A\delta_D(x)^\alpha$. Hence

$$\int_{Q_j} hg^{\varepsilon-1}|\nabla g|^2 dx \ge A^{-1}h(y_j)\delta_D(y_j)^{\alpha(1+\varepsilon)-2}|Q_j| \approx h(y_j)\delta_D(y_j)^{\alpha(1+\varepsilon)-2+n}.$$

Since

$$\left(\int_{Q_j} h^p dx\right)^{1/p} \approx h(y_j)\delta_D(y_j)^{p/n},$$

it follows that if $p < n/(\alpha - 2 + n) \equiv p_k$, then

$$\left(\int_{Q_j} h^p dx\right)^{1/p} \le A \int_{Q_j} hg^{\varepsilon-1}|\nabla g|^2 dx,$$

provided ε is sufficiently small. Summing up the above inequality proves the theorem. □

8.4. Related topics. The proof of Theorem 10 estimates the harmonic measure with the aid of the exterior condition. It could be reasonable to remove the exterior conditions in the context of the previous arguments. Along this direction, [**4**] showed the L^1-integrability of positive superharmonic functions in a John domain with John constant close to 1. If the conical cone condition is valid instead of the twisted cone condition for a John domain, then all positive superharmonic functions are integrable, provided the aperture in the conical cone condition is larger than that corresponding to the Lipschitz condition $k = 1/\sqrt{n-1}$. The L^p-integrability with $p > 1$ for a general John domain remains open. Lindqvist [**55**] showed the integrability of nonlinear superharmonic functions. Gotoh [**47**] studied the integrability of BMO functions on a domain with the quasihyperbolic condition and derived the integrability of nonlinear superharmonic functions. The author [**4**] studied the integrability, relying not on the BHP but on the *Cranston-McConnell inequality* related to *intrinsic ultracontractivity* ([**24**] and [**28**]), the lifetime estimate ([**39**] and [**11**]) and the *perturbation of the Green function* ([**19**] and [**3**]). These subjects have intimate connections to the *parabolic boundary Harnack principle* and the *logarithmic Sobolev inequality*, which cannot be touched in this exposition. Beyond the subject, there seems to be a fascinating and still active research field where real and complex analysis, partial differential equations, differential geometry and probability interact. The author wishes that the survey could shed a light on some work in such an attractive area. The classical study of harmonic functions in a unit ball may lead to the frontier of analysis.

9. Converse of the boundary Harnack principle

Recently, the author [**7**] proved that John domains, uniform domains and inner uniform domains (equivalently, uniformly John domains) are characterized by some potential-theoretic properties. The conditions are necessary and sufficient, provided the domain satisfies the CDC. The 'if parts' of Theorems 12 and 13 below may be regarded as the *converse of the BHP.*

Recall that the CDC holds if and only if there exist constants $\beta > 0$, $A > 1$ and $r_0 > 0$ such that

$$(1) \qquad \omega(x, D \cap S(\xi, r), D \cap B(\xi, r)) \le A \left(\frac{|x-\xi|}{r} \right)^{\beta} \qquad \text{for } x \in D \cap B(\xi, r),$$

whenever $\xi \in \partial D$ and $0 < r < r_0$. This is a uniform upper estimate of the harmonic measure. See also [**6**] for the connection to the Dirichlet problem. A John domain is characterized by a uniform lower estimate.

THEOREM 11. *Let D satisfy the CDC. Then D is a John domain if and only if there exist constants $\alpha > 0$, $A > 1$ and $r_0 > 0$ such that*

$$(2) \qquad \omega(x, D \cap S(\xi, r), D \cap B(\xi, r)) \ge \frac{1}{A} \left(\frac{\delta_D(x)}{r} \right)^{\alpha} \qquad \text{for } x \in D \cap B(\xi, \frac{r}{A}),$$

whenever $\xi \in \partial D$ and $0 < r < r_0$.

A uniform domain is characterized by (2) and the scale invariant BHP, or uniform BHP.

THEOREM 12. *Let D satisfy the CDC. Then D is a uniform domain if and only if the scale invariant BHP and (2) hold.*

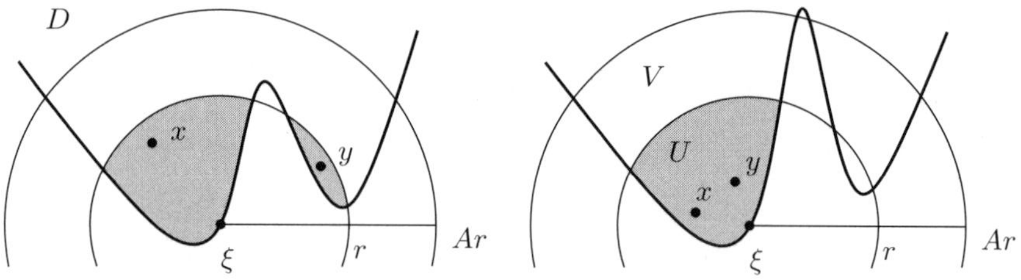

FIGURE 6. Scale invariant BHP and BHP with respect to the internal metric.

An inner uniform domain is characterized by (2) and the uniform BHP with respect to the internal metric. A ball of center at a boundary point with respect to the internal metric becomes a connected component of the intersection of a Euclidean ball and the domain ([**10**, Lemma 2.2]). So, we arrive at the following version of the uniform BHP with respect to the internal metric, which is a property weaker than the uniform BHP. See Figure 6.

DEFINITION. We say that a domain D enjoys the *uniform BHP with respect to the internal metric* or the *scale invariant BHP with respect to the internal metric* if there exist constants $A > 1$ and $r_0 > 0$, depending only on D, with the following property: Let $\xi \in \partial D$, $0 < r < r_0$, U a connected component of $D \cap B(\xi, r)$ and V a connected component of $D \cap B(\xi, Ar)$ including U. Suppose u and v are positive harmonic functions on V vanishing on $\partial D \cap \partial V$. Then

$$\frac{u(x)/u(y)}{v(x)/v(y)} \leq A, \quad \text{whenever } x, y \in U.$$

THEOREM 13. *Let D satisfy the CDC. Then D is an inner uniform domain if and only if the uniform BHP with respect to the internal metric and* (2) *hold.*

Finally, the author acknowledges useful comments by Minoru Murata, Hatano Kaoru, Torbjörn Lundh and the referee.

References

[1] H. Aikawa, *Quasiadditivity of capacity and minimal thinness*, Ann. Acad. Sci. Fenn. Ser. A I Math. **18** no. 1 (1993), 65–75. MR94d:31004

[2] ———, *Integrability of superharmonic functions and subharmonic functions*, Proc. Amer. Math. Soc. **120** no. 1 (1994), 109–117. MR94b:31003

[3] ———, *Norm estimate of Green operator, perturbation of Green function and integrability of superharmonic functions*, Math. Ann. **312** no. 2 (1998), 289–318. MR99m:35042

[4] ———, *Integrability of superharmonic functions in a John domain*, Proc. Amer. Math. Soc. **128** no. 1 (2000), 195–201. MR2000c:31002

[5] ———, *Boundary Harnack principle and Martin boundary for a uniform domain*, J. Math. Soc. Japan **53** no. 1 (2001), 119–145. MR2001m:31007

[6] ———, *Hölder continuity of the Dirichlet solution for a general domain*, Bull. London Math. Soc. **34** no. 6 (2002), 691–702 MR2003f:31007

[7] ———, *Potential theoretic characterizations of nonsmooth domains*, Bull. London Math. Soc. (to appear).

[8] H. Aikawa and M. Essén, *Potential theory—selected topics*, Lecture Notes in Mathematics, vol. 1633, Springer-Verlag, Berlin, 1996. MR98f:31005

[9] H. Aikawa, K. Hirata, and T. Lundh, *Martin boundary points of John domains and unions of convex sets*, preprint.

[10] H. Aikawa, T. Lundh, and T. Mizutani, *Martin boundary of a fractal domain*, Potential Anal. **18** no. 4 (2003), 311–357. MR2004a:31002

[11] H. Aikawa and M. Murata, *Generalized Cranston-McConnell inequalities and Martin boundaries of unbounded domains*, J. Anal. Math. **69** (1996), 137–152. MR98f:31016

[12] A. Ancona, *Principe de Harnack à la frontière et théorème de Fatou pour un opérateur elliptique dans un domaine lipschitzien*, Ann. Inst. Fourier (Grenoble) **28** no. 4 (1978), 169–213. MR80d:31006

[13] ———, *Une propriété de la compactification de Martin d'un domaine euclidien*, Ann. Inst. Fourier (Grenoble) **29** no. 4 (1979), 71–90. MR81f:31013

[14] ———, *Régularité d'accès des bouts et frontière de Martin d'un domaine euclidien*, J. Math. Pures Appl. **63** no. 2 (1984), 215–260. MR86f:31005

[15] ———, *On strong barriers and an inequality of Hardy for domains in* $\mathbf{R}^n$, J. London Math. Soc. (2) **34** no. 2 (1986), 274–290. MR87k:31004

[16] ———, *Negatively curved manifolds, elliptic operators, and the Martin boundary*, Ann. of Math. (2) **125** no. 3 (1987), 495–536. MR88k:58160

[17] ———, *Positive harmonic functions and hyperbolicity*, Potential theory—surveys and problems (Prague, 1987), Springer, Berlin, 1988, pp. 1–23. MR89m:31001

[18] ———, *Sur la frontière de Martin des domaines de Denjoy*, Ann. Acad. Sci. Fenn. Ser. A I Math. **15** no. 2 (1990), 259–271. MR92e:31001

[19] ———, *First eigenvalues and comparison of Green's functions for elliptic operators on manifolds or domains*, J. Anal. Math. **72** (1997), 45–92. MR98i:58212

[20] M. T. Anderson and R. Schoen, *Positive harmonic functions on complete manifolds of negative curvature*, Ann. of Math. (2) **121** no. 3 (1985), 429–461. MR87a:58151

[21] H. Arai, *Boundary behavior of functions on complete manifolds of negative curvature*, Tohoku Math. J. (2) **41** no. 2 (1989), 307–319. MR91c:53037

[22] ———, *Degenerate elliptic operators, Hardy spaces and diffusions on strongly pseudoconvex domains*, Tohoku Math. J. (2) **46** no. 4 (1994), 469–498. MR95h:46034

[23] D. H. Armitage and S. J. Gardiner, *Classical potential theory*, Springer-Verlag London Ltd., London, 2001. MR2001m:31001

[24] R. Bañuelos, *Intrinsic ultracontractivity and eigenfunction estimates for Schrödinger operators*, J. Funct. Anal. **100** no. 1 (1991), 181–206. MR92k:35066

[25] Z. Balogh and A. Volberg, *Boundary Harnack principle for separated semihyperbolic repellers, harmonic measure applications*, Rev. Mat. Iberoamericana **12** no. 2 (1996), 299–336. MR97m:31001

[26] R. F. Bass, *Probabilistic techniques in analysis*, Springer-Verlag, New York, 1995, Correction on http://www.math.uconn.edu/~bass. MR96e:60001

[27] R. F. Bass and K. Burdzy, *A boundary Harnack principle in twisted Hölder domains*, Ann. of Math. (2) **134** no. 2 (1991), 253–276. MR92m:31006

[28] ———, *Lifetimes of conditioned diffusions*, Probab. Theory Related Fields **91** no. 3-4 (1992), 405–443. MR93e:60155

[29] ———, *The Martin boundary in non-Lipschitz domains*, Trans. Amer. Math. Soc. **337** no. 1 (1993), 361–378. MR93g:31019

[30] ———, *Conditioned Brownian motion in planar domains*, Probab. Theory Related Fields **101** no. 4 (1995), 479–493. MR96b:60199

[31] J. Becker and C. Pommerenke, *Hölder continuity of conformal mappings and nonquasiconformal Jordan curves*, Comment. Math. Helv. **57** no. 2 (1982), 221–225. MR84a:30014

[32] M. Benedicks, *Positive harmonic functions vanishing on the boundary of certain domains in* $\mathbf{R}^n$, Ark. Mat. **18** no. 1 (1980), 53–72. MR82h:31004

[33] M. Bonk, J. Heinonen, and P. Koskela, *Uniformizing Gromov hyperbolic spaces*, Astérisque no. 270 (2001), viii+99. MR2003b:30024

[34] M. Brelot and J. L. Doob, *Limites angulaires et limites fines*, Ann. Inst. Fourier (Grenoble) **13** no. 2 (1963), 395–415. MR33:4299

[35] L. Carleson, *On the existence of boundary values for harmonic functions in several variables*, Ark. Mat. **4** (1962), 393–399. MR28:2232

[36] N. Chevallier, *Frontière de Martin d'un domaine de* $\mathbf{R}^n$ *dont le bord est inclus dans une hypersurface lipschitzienne*, Ark. Mat. **27** no. 1 (1989), 29–48. MR91e:31024

[37] C. Constantinescu and A. Cornea, *Ideale Ränder Riemannscher Flächen*, Springer-Verlag, Berlin, 1963. MR28:3151

[38] M. Cranston, E. Fabes, and Z. Zhao, *Conditional gauge and potential theory for the Schrödinger operator*, Trans. Amer. Math. Soc. **307** no. 1 (1988), 171–194. MR90a:60135

[39] M. Cranston and T. R. McConnell, *The lifetime of conditioned Brownian motion*, Z. Wahrsch. Verw. Gebiete **65** no. 1 (1983), 1–11. MR85d:60150

[40] M. C. Cranston and T. S. Salisbury, *Martin boundaries of sectorial domains*, Ark. Mat. **31** no. 1 (1993), 27–49. MR94k:31002

[41] B. E. J. Dahlberg, *Estimates of harmonic measure*, Arch. Rational Mech. Anal. **65** no. 3 (1977), 275–288. MR57:6470

[42] J. L. Doob, *Classical potential theory and its probabilistic counterpart*, Springer-Verlag, Berlin, 2001, Reprint of the 1984 edition. MR2001j:31002

[43] P. Fatou, *Séries trigonométriques et séries de Taylor*, Acta. Math. **30** (1906), 335–400.

[44] S. J. Gardiner, *Minimal harmonic functions on Denjoy domains*, Proc. Amer. Math. Soc. **107** no. 4 (1989), 963–970. MR90c:31013

[45] F. W. Gehring and O. Martio, *Lipschitz classes and quasiconformal mappings*, Ann. Acad. Sci. Fenn. Ser. A I Math. **10** (1985), 203–219. MR87b:30029

[46] F. W. Gehring and B. G. Osgood, *Uniform domains and the quasihyperbolic metric*, J. Analyse Math. **36** (1979), 50–74. MR81k:30023

[47] Y. Gotoh, *Integrability of superharmonic functions, uniform domains, and Hölder domains*, Proc. Amer. Math. Soc. **127** no. 5 (1999), 1443–1451. MR99h:46036

[48] L. L. Helms, *Introduction to potential theory*, Robert E. Krieger Publishing Co., Huntington, N.Y., 1975, Reprint of the 1969 edition, Pure and Applied Mathematics, Vol. XXII, Academic Press, New York. MR57:659

[49] G. Herglotz, *Über Potenzreihen mit positivem, reelen Teil im Einheitskreis*, Ber. Verhandl. Sächs Akad. Wiss. Leipzig, Math.-Phys. Kl. **63** (1911), 501–511.

[50] R. A. Hunt and R. L. Wheeden, *On the boundary values of harmonic functions*, Trans. Amer. Math. Soc. **132** (1968), 307–322. MR37:1634

[51] ———, *Positive harmonic functions on Lipschitz domains*, Trans. Amer. Math. Soc. **147** (1970), 507–527. MR43:547

[52] D. S. Jerison and C. E. Kenig, *Boundary behavior of harmonic functions in nontangentially accessible domains*, Adv. in Math. **46** no. 1 (1982), 80–147. MR84d:31005b

[53] P. W. Jones, *Extension theorems for BMO*, Indiana Univ. Math. J. **29** no. 1 (1980), 41–66. MR81b:42047

[54] ———, *Quasiconformal mappings and extendability of functions in Sobolev spaces*, Acta Math. **147** no. 1-2 (1981), 71–88. MR83i:30014

[55] P. Lindqvist, *Global integrability and degenerate quasilinear elliptic equations*, J. Anal. Math. **61** (1993), 283–292. MR95b:35062

[56] A. Lömker, *Martin boundaries of quasi-sectorial domains*, Potential Anal. **13** no. 1 (2000), 11–67. MR2002k:31023

[57] R. S. Martin, *Minimal positive harmonic functions*, Trans. Amer. Math. Soc. **49** (1941), 137–172. MR2:292h

[58] J. Moser, *On Harnack's theorem for elliptic differential equations*, Comm. Pure Appl. Math. **14** (1961), 577–591. MR28:2356

[59] M. Murata, *Structure of positive solutions to* $(-\Delta + V)u = 0$ *in* $\mathbf{R}^n$, Duke Math. J. **53** no. 4 (1986), 869–943. MR88f:35039

[60] ———, *Martin compactification and asymptotics of Green functions for Schrödinger operators with anisotropic potentials*, Math. Ann. **288** no. 2 (1990), 211–230. MR91m:31006

[61] ———, *On construction of Martin boundaries for second order elliptic equations*, Publ. Res. Inst. Math. Sci. **26** no. 4 (1990), 585–627. MR92d:58216

[62] ———, *Structure of positive solutions of the Schrödinger equation*, Sūgaku **47** no. 4 (1995), 360–376; English transl., Sugaku Expositions **11** no. 1 (1998), 101–121. MR97f:35033

[63] ———, *Martin boundaries of elliptic skew products, semismall perturbations, and fundamental solutions of parabolic equations*, J. Funct. Anal. **194** no. 1 (2002), 53–141. MR2003i:35052

[64] L. Naïm, *Sur le rôle de la frontière de R. S. Martin dans la théorie du potentiel*, Ann. Inst. Fourier (Grenoble) **7** (1957), 183–281. MR20:6608

[65] M. Nakai and T. Tada, *Nonmonotoneity of Picard principle*, Trans. Amer. Math. Soc. **292** no. 2 (1985), 629–644. MR87b:30069

[66] ———, *Extreme nonmonotoneity of the Picard principle*, Math. Ann. **281** no. 2 (1988), 279–293. MR89i:31006

[67] ———, *Monotoneity and homogeneity of Picard dimensions for signed radial densities*, Hokkaido Math. J. **26** no. 2 (1997), 253–296. MR98i:31006

[68] ———, *Picard principle for negative planar potentials*, Math. Ann. **308** no. 2 (1997), 319–332. MR99d:31001

[69] R. Näkki and J. Väisälä, *John disks*, Exposition. Math. **9** no. 1 (1991), 3–43. MR92i:30021

[70] R. G. Pinsky, *Positive harmonic functions and diffusion*, Cambridge University Press, Cambridge, 1995. MR96m:60179

[71] S. Segawa, *Martin boundaries of Denjoy domains*, Proc. Amer. Math. Soc. **103** no. 1 (1988), 177–183. MR89m:31008

[72] ———, *Martin boundaries of Denjoy domains and quasiconformal mappings*, J. Math. Kyoto Univ. **30** no. 2 (1990), 297–316. MR91j:30034

[73] W. Smith and D. A. Stegenga, *Exponential integrability of the quasi-hyperbolic metric on Hölder domains*, Ann. Acad. Sci. Fenn. Ser. A I Math. **16** no. 2 (1991), 345–360. MR93b:30016

[74] D. A. Stegenga and D. C. Ullrich, *Superharmonic functions in Hölder domains*, Rocky Mountain J. Math. **25** no. 4 (1995), 1539–1556. MR97j:31003

[75] J. Väisälä, *Uniform domains*, Tohoku Math. J. (2) **40** no. 1 (1988), 101–118. MR89d:30027

[76] ———, *Relatively and inner uniform domains*, Conform. Geom. Dyn. **2** (1998), 56–88 (electronic). MR99e:30014

[77] J. M. G. Wu, *Comparisons of kernel functions, boundary Harnack principle and relative Fatou theorem on Lipschitz domains*, Ann. Inst. Fourier (Grenoble) **28** no. 4 (1978), 147–167. MR80g:31005

Department of Mathematics, Shimane University, Matsue 690-8504, Japan
E-mail address: `haikawa@math.shimane-u.ac.jp`

Translated by HIROAKI AIKAWA

Amer. Math. Soc. Transl.
(2) Vol. **215**, 2005

Singular integrals and Littlewood-Paley functions

Shuichi Sato

1. Introduction

Let $K \in C^1(\mathbb{R}^n \setminus \{0\})$ be a kernel satisfying

$$|K(x)| \le C|x|^{-n}, \qquad |\nabla K(x)| \le C|x|^{-n-1}; \tag{1.1}$$

$$\int_{a<|x|<b} K(x)\,dx = 0 \qquad \text{for all } a, b \text{ such that } 0 < a < b, \tag{1.2}$$

where $|x| = (x_1^2 + \cdots + x_n^2)^{1/2}$, $x = (x_1, \ldots, x_n)$. Let

$$T(f)(x) = \text{p. v.} \int_{\mathbb{R}^n} f(x-y)K(y)\,dy = \lim_{\epsilon \to 0} \int_{|y|>\epsilon} f(x-y)K(y)\,dy \tag{1.3}$$

be the singular integral operator defined by the kernel K. Then, it is known that T is bounded from $L^p(\mathbb{R}^n)$ to $L^p(\mathbb{R}^n)$ $(1 < p < \infty)$ and from $L^1(\mathbb{R}^n)$ to $L^{1,\infty}(\mathbb{R}^n)$, where $L^{1,\infty}(\mathbb{R}^n)$ is the Lorentz space of all those measurable functions on $\mathbb{R}^n$ that satisfy

$$\|f\|_{L^{1,\infty}} = \sup_{\lambda>0} \lambda\, |\{x \in \mathbb{R}^n : |f(x)| > \lambda\}| < \infty.$$

Here $|\cdot|$ denotes Lebesgue measure. The theory of singular integrals of several variables has been a topic of great interest in analysis since the appearance of the famous papers of Calderón and Zygmund [**3, 4**].

Let $\psi \in \mathcal{S}(\mathbb{R}^n)$ satisfy $\int_{\mathbb{R}^n} \psi(x) = 0$ where $\mathcal{S}(\mathbb{R}^n)$ denotes the Schwartz space of infinitely differentiable, rapidly decreasing functions on $\mathbb{R}^n$. Then, the Littlewood-Paley function is defined by

$$S(f)(x) = S_\psi(f)(x) = \left(\int_0^\infty |f * \psi_t(x)|^2\,dt/t \right)^{1/2},$$

where $\psi_t(x) = t^{-n}\psi(t^{-1}x)$. It is well known that the operator S has mapping properties similar to those of the singular integral T mentioned above.

In this note, we consider singular integrals, Littlewood-Paley functions and oscillatory singular integrals (see Section 4) defined by kernels which are assumed to satisfy certain size and smoothness conditions less restrictive than those stated

---

This article originally appeared in Japanese in Sūgaku **55** (2) (2003), 128–147.
2000 *Mathematics Subject Classification.* Primary 42B25.

above. We shall survey some results on the L^p-L^p boundedness and the L^1-$L^{1,\infty}$ boundedness for those operators, including references to some recent research.

2. Singular integrals and L^p estimates

In the following we assume that $n \geq 2$, unless otherwise noted. First we consider a singular integral in (1.3) which is defined by a homogeneous kernel of the form

$$K(x) = \frac{\Omega(x')}{|x|^n}, \qquad x' = x/|x|.$$

We assume that the function $\Omega \in L^1(S^{n-1})$ satisfies

$$\int_{S^{n-1}} \Omega(x')\, d\sigma(x') = 0, \tag{2.1}$$

where S^{n-1} denotes the unit sphere in $\mathbb{R}^n$ and $d\sigma$ the Lebesgue measure on S^{n-1}. In this case, we also write $T = T_\Omega$. If we denote by $\hat{K}$ the Fourier transform of the principal value distribution defined by the kernel K, then $\hat{K}$ is a homogeneous function of degree 0 and can be expressed as follows:

$$\hat{K}(\xi') = -\int_{S^{n-1}} \Omega(\theta) \left[i\frac{\pi}{2} \operatorname{sgn}(\langle \xi', \theta \rangle) + \log |\langle \xi', \theta \rangle| \right] d\sigma(\theta) \qquad (\xi' \in S^{n-1}), \tag{2.2}$$

where $\langle \cdot, \cdot \rangle$ denotes the inner product in $\mathbb{R}^n$. The definition of the Fourier transform we are using is (for $f \in L^1(\mathbb{R}^n)$)

$$\hat{f}(\xi) = \int_{\mathbb{R}^n} f(x) e^{-2\pi i \langle x, \xi \rangle}\, dx;$$

we also write $\hat{f} = \mathcal{F}(f)$. By using (2.2) we can show that T is bounded on L^2 if $\Omega \in L \log L(S^{n-1})$. Here $L \log L(S^{n-1})$ is the space of all those functions Ω on S^{n-1} that satisfy $\int_{S^{n-1}} |\Omega| \log(2 + |\Omega|)\, d\sigma < \infty$.

Furthermore, by the method of rotations of Calderón and Zygmund (see [**4**]) we can show the following:

(2.3) If $\Omega \in L^1(S^{n-1})$ and Ω is odd, then $T_\Omega : L^p \to L^p$ for all $1 < p < \infty$.

(2.4) If $\Omega \in L \log L(S^{n-1})$, then $T_\Omega : L^p \to L^p$ for all $1 < p < \infty$.

Indeed, when Ω is odd, we can write

$$T_\Omega(f)(x) = \frac{1}{2} \int_{S^{n-1}} \Omega(\theta) \left[\text{p. v.} \int_{-\infty}^{\infty} f(x - t\theta)\, \frac{dt}{t} \right] d\sigma(\theta);$$

therefore, by using the L^p boundedness of the Hilbert transform, we can prove (2.3). Here we recall that the Hilbert transform is a singular integral on $\mathbb{R}$ defined by the principal value integral:

$$H(f)(x) = \text{p. v.} \frac{1}{\pi} \int_{-\infty}^{\infty} f(x - t)\, \frac{dt}{t}.$$

Next, let $\Omega \in L \log L(S^{n-1})$ be an even function. Then, if R_j is the Riesz transform defined by $(R_j f)\hat{}(\xi) = (-i\xi_j/|\xi|)\hat{f}(\xi)$ $(j = 1, 2, \ldots, n)$, we can see that $R_j T_\Omega = T_{\Omega_j}$ for some odd function $\Omega_j \in L^1(S^{n-1})$, and hence, since $I = -\sum_{j=1}^n R_j^2$ (I denotes the identity operator), we have $T_\Omega = -\sum_{j=1}^n R_j T_{\Omega_j}$. Thus we get the L^p boundedness of T_Ω by (2.3) and the L^p boundedness of R_j. To prove (2.4) we write Ω as a sum of an odd function and an even function in $L \log L(S^{n-1})$. Then we apply the above results to the two singular integrals arising from this decomposition of Ω, which completes the proof of (2.4).

We can use the Hardy space $H^1(S^{n-1})$ to improve (2.4). First, we recall the definition of $H^1(S^{n-1})$. The Poisson kernel on the unit sphere S^{n-1} is defined by

$$P_{ry'}(x') = \frac{1}{\omega_{n-1}} \frac{1-r^2}{|ry'-x'|^n}, \qquad 0 \le r < 1,\ x',\ y' \in S^{n-1},$$

where ω_{n-1} is the surface area of S^{n-1}. For $f \in L^1(S^{n-1})$, let $P^+ f(x')$ be the maximal function defined by

$$P^+ f(x') = \sup_{0\le r<1} \left| \int_{S^{n-1}} f(y') P_{rx'}(y')\, d\sigma(y') \right| .$$

The Hardy space $H^1(S^{n-1})$ is the normed space of all those functions $f \in L^1(S^{n-1})$ that satisfy

$$\|f\|_{H^1(S^{n-1})} = \|P^+ f\|_{L^1(S^{n-1})} < \infty.$$

It is known that $L\log L(S^{n-1})$ is a proper subspace of $H^1(S^{n-1})$.

Let $\Omega \in H^1(S^{n-1})$ satisfy (2.1). Then it is known that $T_\Omega : L^p \to L^p$ for all $1 < p < \infty$. We can find this result in Coifman and Weiss [**12**], Connett [**13**] and Ricci and Weiss [**34**]; we can see that this can be proved by applying developed versions of the Calderón-Zygmund method of rotations.

Next, we consider a kernel of the form

$$K(x) = b(|x|)\frac{\Omega(x')}{|x|^n},$$

where, for convenience, we assume b to be bounded. Then the Calderón-Zygmund method of rotations does not work for the singular integral $Tf(x) = \text{p. v.}\, K * f(x)$ defined as in (1.3), as can be seen from the following:

EXAMPLE. Define a function b as

$$b(r) = \begin{cases} \sin r & (r > 1), \\ 0 & (0 \le r \le 1). \end{cases}$$

Assume Ω to be odd. If we try to apply the Calderón-Zygmund method of rotations to the operator T as in (2.3), then the one-dimensional operator which plays the role of the Hilbert transform should be:

$$L(f)(x) = \int_{-\infty}^{\infty} f(x-t) b(|t|)\, dt/t,$$

but a direct computation shows that the Fourier transform $\mathcal{F}\,(b(|t|)/t)\,(\xi)$ is unbounded, and hence L is not bounded on L^2.

When $b \in L^\infty([0,\infty))$ and Ω satisfies a Lipschitz condition, Fefferman [**22**] proved that T is bounded from L^p to L^p for all $1 < p < \infty$. Namazi [**32**] improved this result by replacing the Lipschitz condition by a size condition: $\Omega \in L^q(S^{n-1})$ for some $q > 1$. Under the same assumption, Chen [**7**] proved that the maximal operator T_* is bounded on L^p for all $1 < p < \infty$, where

$$T_*(f)(x) = \sup_{\epsilon>0} \left| \int_{|x-y|>\epsilon} K(x-y) f(y)\, dy \right| .$$

By applying the Littlewood-Paley decomposition and using orthogonality via Fourier transform estimates, Duoandikoetxea and Rubio de Francia [**16**] developed methods which can be used to study various interesting operators in harmonic analysis, including the singular integrals with inhomogeneous kernels mentioned above. Now, we present a result which is a special case of a more general theorem in [**16**]. Let $\{\sigma_k\}_{k=-\infty}^{\infty}$ be a sequence of Borel measures on $\mathbb{R}^n$. Let $\alpha \in (0,1]$. We assume the following:

$$\hat{\sigma}_k(0) = 0; \tag{2.5}$$

$$\|\sigma_k\| = \mathcal{F}(|\sigma_k|)(0) \le c_1, \text{ where } |\sigma_k| \text{ is the total variation of } \sigma_k; \tag{2.6}$$

$$|\hat{\sigma}_k(\xi)| \le c_2|2^k\xi|^{-\alpha}; \tag{2.7}$$

$$|\hat{\sigma}_k(\xi)| \le c_3|2^k\xi|^{\alpha}. \tag{2.8}$$

We further assume that the maximal operator

$$\sigma^*(f)(x) = \sup_k ||\sigma_k| * f(x)|$$

is bounded on L^p for all $p \in (1,\infty)$. Then if we define

$$S(f)(x) = \sum_{k=-\infty}^{\infty} \sigma_k * f(x) \qquad (f \in \mathcal{S}(\mathbb{R}^n)),$$

we have the following result:

THEOREM 1 ([**16**]). *The operator S is bounded on $L^p(\mathbb{R}^n)$ for all $p \in (1,\infty)$.*

As an application of this theorem, we can show the L^p boundedness of T when $\Omega \in L^q(S^{n-1})$ for some $q > 1$ as follows (see [**16**]). Put

$$d\sigma_k = b(|x|)\frac{\Omega(x')}{|x|^n}\chi_{[2^k,2^{k+1})}(|x|)\,dx.$$

Then we have $T(f) = \sum_{k=-\infty}^{\infty} \sigma_k * f$. Define a singular measure τ_r on $\mathbb{R}^n$ which is concentrated on $\{|x| = r\}$ by

$$\langle \tau_r, f\rangle = \int_{S^{n-1}} f(r\theta)\Omega(\theta)\,d\sigma(\theta) \qquad (f \in C_0^\infty(\mathbb{R}^n)),$$

where $C_0^\infty(\mathbb{R}^n)$ denotes the space of all those functions on $\mathbb{R}^n$ that are infinitely differentiable and compactly supported. Then the Schwarz inequality implies

$$|\hat{\sigma}_k(\xi)| \le \|b\|_{L^\infty}(\log 2)^{1/2}\left(\int_{2^k}^{2^{k+1}} |\hat{\tau}_r(\xi)|^2\,dr/r\right)^{1/2}.$$

By a straightforward computation involving Hölder's inequality, we have

$$\begin{aligned}
&\int_{2^k}^{2^{k+1}} |\hat{\tau}_r(\xi)|^2\,dr/r \\
&\quad = \iint \Omega(\theta)\bar{\Omega}(\omega)\left(\int_{2^k}^{2^{k+1}} \exp\left(-2\pi i r\langle\xi,\theta-\omega\rangle\right)dr/r\right)d\sigma(\theta)\,d\sigma(\omega) \\
&\quad \le C|2^k\xi|^{-1/(2q')}\|\Omega\|_{L^q(S^{n-1})}^2\left(\iint_{S^{n-1}\times S^{n-1}} |\langle\xi',\theta-\omega\rangle|^{-1/2}\,d\sigma(\theta)\,d\sigma(\omega)\right)^{1/q'},
\end{aligned}$$

where q' denotes the conjugate exponent to q. Since we can easily see that the last integral is bounded in $\xi' \in S^{n-1}$, we have (2.7) with $\alpha = 1/(4q')$. Also, by using (2.1) we can see that $|\hat{\sigma}_k(\xi)| \le C|2^k\xi|$. By using these estimates we can see that the conditions (2.5)–(2.8) hold. Moreover, since $\Omega \in L^1(S^{n-1})$, σ^* is bounded on L^p for all $p \in (1,\infty)$. Consequently, Theorem 1 implies the L^p boundedness of T.

It is known that this theory can be used to prove the L^p boundedness of T even in the case when $\Omega \in L\log L(S^{n-1})$. Fan and Pan [**18, 19**] further relaxed the condition imposed on Ω as follows:

THEOREM 2. *Suppose that $\Omega \in H^1(S^{n-1})$ and that Ω satisfies* (2.1). *Let $b \in L^\infty$. Then the operator T is bounded on $L^p(\mathbb{R}^n)$ for all $p \in (1,\infty)$.*

An atomic decomposition of $\Omega \in H^1(S^{n-1})$ can be used to prove this theorem. Here we recall that a function $a(x')$ is called a q-atom $(1 < q \le \infty)$ if

$$a \in L^q(S^{n-1}), \qquad \int_{S^{n-1}} a(y')\,d\sigma(y') = 0,$$

and, for some $x_0' \in S^{n-1}$ and some $\rho \in (0,1]$,

$$\begin{aligned}\operatorname{supp}(a) \subset \{y' \in S^{n-1} : |y' - x_0'| < \rho\},\\ \|a\|_q \le \rho^{(n-1)(1/q-1)}.\end{aligned}$$

It is known that any $\Omega \in H^1(S^{n-1})$ that satisfies (2.1) admits an atomic decomposition $\Omega = \sum \lambda_j a_j$, where each a_j is a q-atom and $\sum |\lambda_j| \le C\|\Omega\|_{H^1(S^{n-1})}$. Fan and Pan further developed the theory of [**16**] mentioned above so that they could prove the estimates $\|T(f)\|_p \le C\|f\|_p$ when Ω is an atom, with a constant C independent of an atom. By using this, they proved Theorem 2. For the proof, the linear transformation

$$(x_1, x_2, \dots, x_n) \mapsto (\rho^2 x_1, \rho x_2, \dots, \rho x_n) \qquad (\rho > 0),$$

which is closely related to the geometry of S^{n-1}, can be effectively used. This method also applies to the maximal operator T_*.

REMARK 1. We have proved above the L^p boundedness of a singular integral T by applying Theorem 1, where σ_k was a function (an absolutely continuous measure). But when we treat singular integrals associated with curves or surfaces, we need to take suitable singular measures σ_k and further develop the theory (see [**16**]).

REMARK 2. In their study of singular integrals with homogeneous kernel $\Omega(x')/|x|^n$, paying attention to the appearance of $\log|\langle \xi', \theta\rangle|$ in (2.2), Grafakos and Stefanov [**23**] introduced the condition

$$\sup_{\xi' \in S^{n-1}} \int_{S^{n-1}} |\Omega(\theta)| \left(\log \frac{1}{|\langle \xi', \theta\rangle|}\right)^{1+\epsilon} d\sigma(\theta) < \infty,$$

where $\epsilon > 0$. Assuming this condition, they proved the L^p boundedness of T_Ω, where the range of p (contained in $(1,\infty)$) depends on ϵ.

3. Singular integrals and weak type $(1,1)$ estimates

Let T be a translation-invariant operator that is bounded on L^2. Suppose that there exists a kernel K that is locally integrable in $\mathbb{R}^n \setminus \{0\}$ and for which we have

$$\langle T(f), g\rangle = \iint_{\mathbb{R}^n\times\mathbb{R}^n} g(x)f(y)K(x-y)\,dx\,dy$$

for all $f, g \in C_0^\infty(\mathbb{R}^n)$ such that $\operatorname{supp}(f)$ and $\operatorname{supp}(g)$ are disjoint. If we assume that K satisfies the Hörmander condition

$$\sup_{y\in\mathbb{R}^n}\int_{|x|>2|y|} |K(x-y)-K(x)|\,dx < \infty,$$

then T is of weak type $(1,1)$ (see [**26**], [**37**]). This condition will be satisfied if the kernel K has the property (1.1).

Here we see some examples of the weak type $(1,1)$ operators whose kernels do not necessarily satisfy the Hörmander condition. For a kernel K, define

$$V(\theta) = \sup_{R>0} V_R(\theta), \qquad \text{where} \quad V_R(\theta) = \int_R^{2R} |K(r\theta)|r^{n-1}\,dr.$$

For $t \in (0,1]$, $q > 0$, put

$$\omega_q(t) = \sup_{|s|<tR/2}\left(R^{-n}\int_{S^{n-1}}\int_R^{2R} |R^n[K((r-s)\theta)-K(r\theta)]|^q r^{n-1}\,dr\,d\sigma(\theta)\right)^{1/q},$$

where the supremum is taken over all s, R satisfying $|s| < tR/2$. Also, when $q=\infty$, by the usual modification we can define $\omega_\infty(t)$. (Here we need ω_q only for $q=1$; ω_q, $q>1$, can be used to get the weighted weak type $(1,1)$ estimates.) The following result is due to Seeger [**48**] (see also Tao [**56**] and Seeger and Tao [**49**] for further developments).

THEOREM 3 (A. Seeger). *Let T be bounded on $L^2(\mathbb{R}^n)$. Suppose the following:*
(1) $V \in L\log L(S^{n-1})$,
(2) $\int_0^1 \omega_1(t)\,dt/t < \infty$.
Then T is of weak type $(1,1)$.

Now we consider a kernel of the form

$$K(x) = |x|^{-n}h(|x|)\Omega(x'), \tag{3.1}$$

where $\Omega \in L\log L(S^{n-1})$ is assumed to satisfy $\int \Omega(\theta)\,d\sigma(\theta) = 0$ (the condition (2.1)). Also, for h we assume that

$$\sup_{R>0}\int_R^{2R} |h(r)|\,dr/r < \infty$$

and $\int_0^1 \omega_1(h,t)\,dt/t < \infty$, where, for $0 < t \le 1$,

$$\omega_1(h,t) = \sup_{|s|<Rt/2}\int_R^{2R} |h(u-s)-h(u)|\,du/u.$$

Then, since it is known that the operator $Tf = \text{p.v.}\,K * f$ is bounded on $L^2(\mathbb{R}^n)$, Theorem 3 implies the weak type $(1,1)$ boundedness of T. When $h \equiv 1$, the lower-dimensional case of this result is in Christ and Rubio de Francia [**9**]. When $n=2$,

$h \equiv 1$ and $\Omega \in L^q$ for some $q > 1$, the weak type $(1,1)$ boundedness is in Hofmann [**24**] (see also [**25**]).

Next we see results for operators with a kernel of the form (3.1), where no smoothness condition is assumed for the function h. Define the L^q-modulus of continuity of Ω as

$$\omega_q(\Omega, t) = \sup_{|\rho|<t} \left(\int_{S^{n-1}} |\Omega(\rho\theta) - \Omega(\theta)|^q \, d\sigma(\theta) \right)^{1/q},$$

where ρ is a rotation and $|\rho| = \sup_\theta |\theta - \rho\theta|$.

THEOREM 4. *Suppose T is bounded on $L^2(\mathbb{R}^n)$. We further assume the following*:

(1) *the function h is bounded,*

(2) $\int_0^1 \omega_1(\Omega, t)\, dt/t < \infty$.

Then T is of weak type $(1,1)$.

An important special case of this result, where the condition (2) is replaced by a Lipschitz condition, is due to Christ and Rubio de Francia [**9**]. Theorem 4 in full generality is essentially shown by Sun [**55**] (see also [**41**]). Here we note that the condition (2.1) ($\int \Omega(\theta)\, d\sigma(\theta) = 0$) is not assumed in Theorem 4.

We recall that the Bochner-Riesz mean of order δ on $\mathbb{R}^n$ is defined by

$$S_R^\delta(f)(x) = \int_{|\xi|<R} \hat{f}(\xi)(1 - R^{-2}|\xi|^2)^\delta \, e^{2\pi i x\xi} \, d\xi = H_{R^{-1}}^\delta * f(x), \tag{3.2}$$

where

$$H^\delta(x) = \pi^{-\delta}\Gamma(\delta+1)|x|^{-(n/2+\delta)} J_{n/2+\delta}(2\pi|x|).$$

Here J_ν denotes the Bessel function of the first kind of order ν. If $\alpha = (n-1)/2$, then S_1^α is bounded on L^2 and $H^\alpha(x) \le c(1+|x|)^{-n}$. Therefore, taking $h(r) = r^n H^\alpha(r)$ and $\Omega \equiv 1$ in Theorem 4, we get the following theorem of Christ [**8**] (see also Christ and Sogge [**10**]).

THEOREM 5. *S_1^α is of weak type* $(1,1)$.

Christ [**8**] has further developed the Calderón-Zygmund methods to get the weak $(1,1)$ boundedness of operators whose kernels do not necessarily satisfy the Hörmander condition. We can find in [**8**] the basic scheme for the proofs of the above theorems on the weak $(1,1)$ boundedness. We shall see more about that scheme later in Section 4 when we study oscillatory singular integrals, where we shall also see the computations involving the kernels that appear in the proof of the special case of Theorem 4.

Let w be a measurable function such that $w(x) > 0$ a.e. (a weight function). We say $w \in A_1$ if there exists a non-negative constant C such that $M(w)(x) \le C\, w(x)$ a.e. Here M denotes the Hardy-Littlewood maximal operator:

$$M(w)(x) = \sup_{x \in Q} |Q|^{-1} \int_Q w(x)\, dx,$$

where the supremum is taken over all cubes Q in $\mathbb{R}^n$ such that $x \in Q$. The weighted space L_w^p is the vector space of all measurable functions f such that

$$\|f\|_{L_w^p} = \left(\int_{\mathbb{R}^n} |f(x)|^p w(x)\, dx \right)^{1/p} < \infty.$$

Also, $L_w^{p,\infty}$ denotes the space of all those measurable functions f that satisfy

$$\|f\|_{L_w^{p,\infty}} = \sup_{\lambda>0} \lambda w\left(\{x \in \mathbb{R}^n : |f(x)| > \lambda\}\right)^{1/p} < \infty,$$

where $w(E) = \int_E w(x)\,dx$. By Vargas [**57**], Theorem 5 has been extended to the weighted Lebesgue spaces as follows.

THEOREM 6. *Let $w \in A_1$. Then*

$$\|S_1^\alpha(f)\|_{L_w^{1,\infty}} \le C\|f\|_{L_w^1}.$$

Furthermore, it is known that Theorem 5 extends to vector-valued inequalities on weighted Lebesgue spaces as follows:

THEOREM 7. *Let $w \in A_1$. Then for every sequence $\{R_k\}_{k=1}^\infty$ of positive numbers we have*

$$\left\| \left(\sum_{k=1}^\infty |S_{R_k}^\alpha(f_k)|^2 \right)^{1/2} \right\|_{L_w^{1,\infty}} \le C \left\| \left(\sum_{k=1}^\infty |f_k|^2 \right)^{1/2} \right\|_{L_w^1}.$$

This was shown step by step in [**38**], [**39**], [**41**] (see also Seeger [**47**]). As an application of Theorem 7, we have $\lim_{k\to\infty} S_{2^k}^\alpha(f)(x) = f(x)$ a.e. for $f \in H_w^1$ ($w \in A_1$), where H_w^1 is the weighted Hardy space. The divergence of $S_R^\alpha(f)$ on H_w^1 is studied in [**40**].

When $\Omega \in L^1(S^{n-1})$ is odd, as we mentioned in Section 2, the singular integral T_Ω is bounded on L^p for all $1 < p < \infty$. However, the question whether T_Ω can be of weak type $(1,1)$ is yet to be settled; this is also the case with T_Ω for $\Omega \in H^1(S^{n-1})$. When the kernel K has the form (3.1), to apply Theorem 3 we need a certain smoothness condition for h and to apply Theorem 4 we need one for Ω. It will be interesting to study the question whether these results can be extended to the case when both h and Ω are rough.

4. Oscillatory singular integrals

Let $P(x,y)$ be a real-valued polynomial on $\mathbb{R}^n \times \mathbb{R}^n$:

$$P(x,y) = \sum_{|\alpha|\le M, |\beta|\le N} a_{\alpha\beta} x^\alpha y^\beta, \tag{4.1}$$

where $\alpha = (\alpha_1, \dots, \alpha_n)$ is a multi-index of non-negative integers, $|\alpha| = \alpha_1 + \cdots + \alpha_n$ and $x^\alpha = x_1^{\alpha_1} \dots x_n^{\alpha_n}$. Suppose a kernel K satisfies (1.1) and (1.2). Define an oscillatory singular integral T by

$$T(f)(x) = \text{p. v.} \int_{\mathbb{R}^n} e^{iP(x,y)} K(x-y) f(y)\,dy = \lim_{\epsilon\to 0} \int_{|x-y|>\epsilon} e^{iP(x,y)} K(x-y) f(y)\,dy.$$

Then we have the following (Theorems 8, 9, 10):

THEOREM 8 (Ricci-Stein [**33**]). *Let $1 < p < \infty$. Then T is bounded on $L^p(\mathbb{R}^n)$:*

$$\|T(f)\|_{L^p} \le C\|f\|_{L^p},$$

where the constant C is independent of a polynomial if the polynomial's degree is fixed (in particular, C is independent of the size of the polynomial coefficients).

THEOREM 9 (Chanillo-Christ [**6**]). *The operator T is bounded from $L^1(\mathbb{R}^n)$ to $L^{1,\infty}(\mathbb{R}^n)$:*

$$\|T(f)\|_{L^{1,\infty}} \le C\|f\|_{L^1},$$

where the constant C is independent of a polynomial if the polynomial is of fixed degree.

THEOREM 10 ([**43**]). *Let $w \in A_1$. Then $T : L^1_w \to L^{1,\infty}_w$; that is,*

$$\|T(f)\|_{L^{1,\infty}_w} \le C\|f\|_{L^1_w},$$

where the constant C is independent of a polynomial if the polynomial is of fixed degree.

Let K be a locally integrable function on $\mathbb{R}^n \setminus \{0\}$. For $r \ge 1$, $0 < t \le 1$, $R > 0$, put

$$\omega_{r,R}(t) = \sup_{|y|\le Rt/2} \left(R^{-n} \int_{R\le|x|\le 2R} |R^n (K(x-y) - K(x))|^r \, dx \right)^{1/r}.$$

We say that K satisfies the D_r-condition if

$$B_r = \int_0^1 \omega_r(t)\, dt/t < \infty, \qquad \text{where} \quad \omega_r(t) = \sup_{R>0} \omega_{r,R}(t);$$

$$C_r = \sup_{R>0} \left(R^{-n} \int_{R\le|x|\le 2R} |R^n K(x)|^r \, dx \right)^{1/r} < \infty.$$

By a suitable modification, we can also define the D_∞-condition. Theorems 9 and 10 can be improved in the following manner:

THEOREM 11. *Let $r > 1$, $1/r + 1/u = 1$, and let $w^u \in A_1$. Suppose that the kernel K satisfies the D_r-condition and* (1.2). *Then we have*

$$\|T(f)\|_{L^{1,\infty}_w} \le C\|f\|_{L^1_w},$$

where the constant C is independent of a polynomial if we fix the polynomial's degree.

THEOREM 12. *If the kernel K satisfies the D_1-condition and* (1.2), *then*

$$\|T(f)\|_{L^{1,\infty}} \le C\|f\|_{L^1},$$

where the constant C is independent of a polynomial if the polynomial is of fixed degree.

These results are shown by double induction with respect to the degrees in x and y of the polynomial. In the following, we shall briefly show the proof of Theorem 11, concentrating on a result (Proposition 1) which plays an important role in the induction process in proving Theorem 11; we can prove Theorem 12 in the same way. We write a polynomial P as in (4.1). We assume that there exist α and β such that $|\alpha| = M$, $a_{\alpha\beta} \neq 0$. We write the polynomial as

$$P(x,y) = \sum_{|\alpha|\le M} x^\alpha Q_\alpha(y)$$

and put $L = \max\{\deg(Q_\alpha) : Q_\alpha \neq 0, |\alpha| = M\}$. Then we have $0 \leq L \leq N$. We assume that $L \geq 1$ and $\max_{|\alpha|=M,|\beta|=L} |a_{\alpha\beta}| = 1$. Under these assumptions on the polynomial P, we define

$$T_\infty(f)(x) = \int_{|x-y|>1} e^{iP(x,y)} K(x-y) f(y)\, dy.$$

We shall apply the following result with suitable M, N and L in the induction process in proving Theorem 11:

PROPOSITION 1. *Let a kernel K and a weight w be as in Theorem 11. Then we have*

$$\|T_\infty(f)\|_{L^{1,\infty}_w} \leq C\|f\|_{L^1_w},$$

where the constant C is independent of a polynomial if we fix the polynomial's degree.

We can employ the procedure of Christ [**8**] as the basic framework for the proof of Proposition 1. This is possible because we can show the key estimates (see Lemma 2) which occur in the unweighted computation involving the Calderón-Zygmund decomposition. To prove Lemma 2 we need a geometrical result for polynomials (Lemma 5).

For $f \in \mathcal{S}(\mathbb{R}^n)$ and $\lambda > 0$, by the Calderón-Zygmund decomposition we have a collection of non-overlapping dyadic cubes $\{Q\}$ and functions g, b such that

$$f = g + b;$$
$$c_1\lambda \leq |Q|^{-1}\int_Q |f| \leq c_2\lambda, \quad v\Big(\bigcup Q\Big) \leq c_3\|f\|_{L^1(M(v))}/\lambda;$$
$$\|g\|_\infty \leq c_4\lambda, \quad \|g\|_{L^1_v} \leq c_5\|f\|_{L^1(M(v))};$$
$$b = \sum_Q b_Q, \quad \operatorname{supp}(b_Q) \subset Q, \quad \int b_Q = 0, \quad \|b_Q\|_{L^1} \leq c_6\lambda|Q|,$$

where v is a weight function and $c_1, \ldots, c_6$ are positive constants. We can treat $T_\infty(g)$ by the ordinary Calderón-Zygmund theory via the L^2_w estimates for T_∞, which is easier to show.

Let

$$B_i = \sum_{|Q|=2^{in}} b_Q \quad (i \geq 1), \qquad B_0 = \sum_{|Q|\leq 1} b_Q.$$

Choose $\varphi \in C_0^\infty(\mathbb{R})$ such that

$$\operatorname{supp}(\varphi) \subset \{1/2 \leq r \leq 2\}, \qquad \sum_{j=-\infty}^{\infty} \varphi(2^{-j}r) = 1 \quad (r \neq 0),$$

and define

$$T_j(f)(x) = \int e^{iP(x,y)} K(x-y)\varphi(2^{-j}|x-y|) f(y)\, dy.$$

Put $S(f) = \sum_{j=1}^\infty T_j(f)$.

Put $E = \bigcup Q^*$, where Q^* is a cube with the same center as Q and side length 100 times that of Q (we assume the sides of Q^* to be parallel to the coordinate axes). Then we note that $w(E) \le C\lambda^{-1}\|f\|_{L^1_w}$. The principal part of the proof of Proposition 1 is to show that

$$w\left(\{x \in \mathbb{R}^n \setminus E : |S(b)(x)| > \lambda\}\right) \le C\lambda^{-1}\|f\|_{L^1_w}. \tag{4.2}$$

Put $K_j(x,y) = e^{iP(x,y)}K(x-y)\varphi(2^{-j}|x-y|)$. If $x \in \mathbb{R}^n \setminus E$, taking into account the support conditions of the kernels and b_Q, we see that $S(b)(x)$ is equal to

$$\sum_{i\ge 0}\sum_{j\ge 1}\int K_j(x,y)B_i(y)\,dy = \sum_{s\ge 1}\sum_{j\ge s}\int K_j(x,y)B_{j-s}(y)\,dy = \sum_{s\ge 1}\sum_{j\ge s}T_j(B_{j-s})(x).$$

Therefore to get (4.2) it suffices to show that

$$w\left(\left\{x \in \mathbb{R}^n : \left|\sum_{s\ge 1}\sum_{j\ge s}T_j(B_{j-s})(x)\right| > \lambda\right\}\right) \le C\lambda^{-1}\|f\|_{L^1_w}. \tag{4.3}$$

One of the main ingredients of the proof of (4.3) is the following:

LEMMA 1. *Let $\{L_j\}_{j\ge 1}$ be a sequence of functions such that*

$$\operatorname{supp}(L_j) \subset \{2^{j-6} \le |x| \le 2^{j+6}\}, \quad |L_j(x)| \le c|x|^{-n}, \quad |\nabla L_j(x)| \le c|x|^{-n-1}.$$

Define

$$G_j(f)(x) = \int_{\mathbb{R}^n} e^{iP(x,y)}L_j(x-y)f(y)\,dy.$$

Let $w \in A_1$. Then there exists an $\epsilon > 0$ such that for every integer $s \ge 1$,

$$w\left(\left\{x \in \mathbb{R}^n : \left|\sum_{j\ge s}G_j(B_{j-s})(x)\right| > \lambda\right\}\right) \le C2^{-\epsilon s}\lambda^{-1}\|f\|_{L^1_w}.$$

If we assume this lemma, (4.3) can be proved as follows. First we decompose

$$T_j(B_{j-s}) = \int e^{iP(x,y)}L_j^s(x-y)B_{j-s}(y)\,dy + \int e^{iP(x,y)}R_j^s(x-y)B_{j-s}(y)\,dy$$
$$=: G_j^s(B_{j-s})(x) + H_j^s(B_{j-s})(x).$$

Here L_j^s satisfies the hypotheses on L_j in Lemma 1 with the constants depending on s; the constants are so suitably depending on s that by using Lemma 1 we can obtain

$$w\left(\left\{x \in \mathbb{R}^n : \left|\sum_{s\ge 1}\sum_{j\ge s}G_j^s(B_{j-s})(x)\right| > \lambda\right\}\right) \le C\lambda^{-1}\|f\|_{L^1_w}$$

(see [**43**], [**44**] for the details). Furthermore, we can use the Dini condition and the assumption on the weight to get the desired estimates for $H_j^s(B_{j-s})$. Combining these results, we can get (4.3).

To prove Lemma 1 we need Lemmas 2 and 3 below.

LEMMA 2. *Let G_j be as in Lemma 1. Then for every positive integer s we have*

$$\left\|\sum_{j\ge s}G_j(B_{j-s})\right\|_{L^2}^2 \le C\lambda 2^{-s}\sum_{j\ge 0}\|B_j\|_{L^1}.$$

LEMMA 3. *Let v be a locally integrable, non-negative function. Then for every positive integer s we have*

$$\left\| \sum_{j \geq s} G_j(B_{j-s}) \right\|_{L^1_v} \leq C\lambda \sum_Q |Q| \inf_{x \in Q} M(v)(x).$$

By using a modified version of the interpolation method of A. Vargas [**57**] between the estimates in Lemmas 2 and 3, we can get Lemma 1. The proof of Lemma 3 is easy. We now outline the proof of Lemma 2. For integers $k, m \geq 1$, put

$$H_{km}(x,y) = \int e^{-iP(z,x)+iP(z,y)} \overline{L}_k(z-x) L_m(z-y)\, dz. \tag{4.4}$$

Then we note that $G_k^* G_m(f)(x) = \int H_{km}(x,y) f(y)\, dy$, where G_k^* is the adjoint of G_k.

LEMMA 4. *Let j, s be integers such that $0 < s \leq j$. Then we have*

$$\sum_{i=s}^{j} \sup_{x \in \mathbb{R}^n} \left| \int B_{i-s}(y) H_{ji}(x,y)\, dy \right| \leq C\lambda 2^{-s}.$$

Let $\langle \cdot, \cdot \rangle$ denote the inner product in $L^2(\mathbb{R}^n)$. If we assume Lemma 4, then Lemma 2 can be proved as follows:

$$\begin{aligned}
\left\| \sum_{j \geq s} G_j(B_{j-s}) \right\|_{L^2}^2 &\leq 2 \sum_{j \geq s} \sum_{i=s}^{j} |\langle G_j(B_{j-s}), G_i(B_{i-s}) \rangle| \\
&\leq 2 \sum_{j \geq s} \sum_{i=s}^{j} |\langle B_{j-s}, G_j^* G_i(B_{i-s}) \rangle| \\
&\leq 2 \sum_{j \geq s} \|B_{j-s}\|_{L^1} \left[\sum_{i=s}^{j} \|G_j^* G_i(B_{i-s})\|_{L^\infty} \right] \\
&\leq C\lambda 2^{-s} \sum_{j \geq s} \|B_{j-s}\|_{L^1}.
\end{aligned}$$

To prove Lemma 4 we use a result (Lemma 5) which can be shown by algebraic geometry methods. In the following, we denote by $P(x)$ a real-valued polynomial on $\mathbb{R}^n$.

DEFINITION 1. For a polynomial $P(x) = \sum_{|\alpha| \leq N} a_\alpha x^\alpha$ of degree N, put $\|P\| = \max_{|\alpha|=N} |a_\alpha|$.

DEFINITION 2. For $\beta > 0$, put $\mathcal{R}(P, \beta) = \{x \in \mathbb{R}^n : |P(x)| \leq \beta\}$.

LEMMA 5 ([**43**]). *Let $d(E,F)$ denote the distance between the sets E and F. Let $R \geq s > 0$, $\gamma > 0$. Suppose that* $\deg(P) = N \geq 1$, $\|P\| = 1$. *Let $a \in \mathbb{R}^n$. Then*

$$\left| \{x \in \mathbb{R}^n : |a - x| \leq R, d\left(x, \mathcal{R}(P, s^N)\right) \leq \gamma s\} \right| \leq C_{n,N,\gamma} R^{n-1} s,$$

where $C_{n,N,\gamma}$ is a non-negative constant independent of $a \in \mathbb{R}^n$.

This can be proved by applying the algebraic geometry argument of Chanillo and Christ [**6**]. Now we can give a rough sketch of the proof of Lemma 4. First we write

$$\exp\left(i[P(z,y)-P(z,x)]\right)=\exp\left(i\left[\sum_{|\alpha|\leq M} z^{\alpha}(Q_\alpha(y)-Q_\alpha(x))\right]\right)$$

in (4.4) and apply van der Corput's lemma with respect to the z variable to estimate the kernel H_{km}. Then using the kernel estimates and Lemma 5, we estimate the integral $\int B_{i-s}(y)H_{ji}(x,y)\,dy$; by suitably applying Lemma 5 to the polynomial $Q_\alpha(y)-Q_\alpha(x)$ ($|\alpha|=M$) in the y variable (fixing x and recalling the assumption on the polynomial in Proposition 1), we can get those estimates for the integral that can be used to reach the conclusion of Lemma 4 (see [**43**], [**44**] for the rigorous details).

Lu and Zhang [**31**] extended Theorem 8 to the case where the kernel has the form $K(x)=|x|^{-n}\Omega(x')$ with Ω satisfying $\Omega\in L^q(S^{n-1})$ for some $1<q\leq\infty$ and $\int\Omega\,d\sigma=0$. It will be interesting to study the question whether Theorem 9 can also be extended to the same setting.

The scheme for the proof of Theorem 4 is basically similar to that for Proposition 1. To conclude this section, we consider the case $K(x)=h(|x|)/|x|^n$ in Theorem 4. Put $K_j(x)=\varphi(2^{-j}|x|)K(x)$ and $B_i=\sum_{|Q|=2^{in}}b_Q$ for $i,j\in\mathbb{Z}$ (the set of all integers). Then a principal part of the proof of Theorem 4 is to show that there exists an $\epsilon>0$ such that

$$\|\sum_{j\in\mathbb{Z}}K_j*B_{j-s}\|_{L^2}^2\leq C2^{-\epsilon s}\lambda\|f\|_{L^1}\qquad(s\geq 1). \tag{4.5}$$

This should be compared with Lemma 2. Put $\tilde{K}_j(x)=\bar{K}_j(-x)$. We can prove (4.5) by using the following estimates in the same way as we proved Lemma 2 by using Lemma 4: for every $j\in\mathbb{Z}$ we have

$$\|\tilde{K}_j*K_j*B_{j-s}\|_{L^\infty}\leq C\lambda 2^{-\epsilon s}, \tag{4.6}$$

$$\sum_{i\leq j-10}\|\tilde{K}_j*K_i*B_{i-s}\|_{L^\infty}\leq C\lambda 2^{-\epsilon s}, \tag{4.7}$$

where $\epsilon>0$ is a positive number independent of $j\in\mathbb{Z}$. To show (4.6) and (4.7) we can use the following estimates (see [**9**], [**38**], [**39**]):

LEMMA 6. *Suppose that $x\in\mathbb{R}^n\setminus\{0\}$ and $|y|<|x|/2$. Then we have the following*:

(1) $$|\tilde{K}_0*K_j(x+y)-\tilde{K}_0*K_j(x)|\leq C|2^{-j}y|^{1/2}\quad(j\leq -10),$$

(2) $$|\tilde{K}_0*K_0(x+y)-\tilde{K}_0*K_0(x)|\leq C|y|^{1/2}|x|^{-3/2}.$$

We can obtain these results since, even if the functions K_j are difficult to handle, by considering convolution between them we get functions that are easier to estimate. Indeed, we have the following equality:

$$\tilde{K}_0*K_j(x)=c\int_0^\infty\int_0^\infty \bar{h}(r)h(s)[\sigma_r*\sigma_s(x)]\varphi(r)\varphi(2^{-j}s)\,\frac{dr}{r}\,\frac{ds}{s}, \tag{4.8}$$

where σ_r denotes the uniform probability measure on $\{x : |x| = r\}$. Let $r \geq s$. Then we note that if $r - s < |x| < r + s$, then

$$\sigma_r * \sigma_s(x) = c_n r^{-n+2} s^{-n+2} |x|^{-n+2} \left[((r+s)^2 - |x|^2)(|x|^2 - (r-s)^2)\right]^{(n-3)/2},$$

and $\sigma_r * \sigma_s(x) = 0$ otherwise. By using this in (4.8), we can prove Lemma 6.

5. Littlewood-Paley functions

Let $\mathbb{T}$ be the one-dimensional torus, which is identified with $\mathbb{R}/\mathbb{Z}$. For $f \in L^p(\mathbb{T})$ $(1 < p < \infty)$, let $\sum_{k=-\infty}^{\infty} c_k e^{2\pi i k\theta}$, $c_k = \int_{\mathbb{T}} f(x) e^{-2\pi i k x}\, dx$, be its Fourier series. For a positive integer m, put

$$\Delta_m(\theta) = \sum_{2^{m-1} \leq |k| < 2^m} c_k e^{2\pi i k\theta};$$

also put $\Delta_0 = c_0$, and define

$$\gamma(f) = \left(\sum_{m=0}^{\infty} |\Delta_m|^2\right)^{1/2}.$$

Then Littlewood and Paley [**28, 29, 30**] proved that there exist positive constants A_p and B_p such that

$$A_p \|f\|_{L^p} \leq \|\gamma(f)\|_{L^p} \leq B_p \|f\|_{L^p}. \tag{5.1}$$

$\gamma(f)$ is called the Littlewood-Paley function. (5.1) can be applied to obtain results such as the multiplier theorems of Marcinkiewicz type and the lacunary convergence of the Fourier series. Furthermore, Littlewood and Paley [**28, 29, 30**] proved a result analogous to (5.1) for the function g defined by

$$g(f)(\theta) = \left(\int_0^1 (1-t)|(\partial/\partial t)P_t * f(\theta)|^2\, dt\right)^{1/2},$$

where

$$P_t(\theta) = \frac{1 - t^2}{1 - 2t\cos(2\pi\theta) + t^2} \qquad \text{(the Poisson kernel)},$$

and also studied the relation between $\gamma(f)$ and $g(f)$ (see also Zygmund [**Chap. XV, 59**]).

For a function $\psi \in L^1(\mathbb{R}^n)$ satisfying

$$\int_{\mathbb{R}^n} \psi(x)\, dx = 0, \tag{5.2}$$

the Littlewood-Paley function on $\mathbb{R}^n$ is defined by

$$S(f)(x) = S_\psi(f)(x) = \left(\int_0^\infty |\psi_t * f(x)|^2\, dt/t\right)^{1/2}.$$

Refer to [**51, 52, 54**] as well for the history and significance of the Littlewood-Paley function in harmonic analysis. Also see [**15, 20, 21, 42**] for recent developments of the theory. For the L^p boundedness $(1 < p < \infty)$ of S_ψ, the following result (valid still in the case $n = 1$) of Benedek, Calderón and Panzone [**1**] is well known.

THEOREM 13. *Suppose that* ψ *satisfies* (5.2). *Moreover, suppose that there exists a positive number* ϵ *such that*

$$|\psi(x)| \le c(1+|x|)^{-n-\epsilon}, \tag{5.3}$$

$$\int_{\mathbb{R}^n} |\psi(x-y)-\psi(x)|\,dx \le c|y|^{\epsilon}. \tag{5.4}$$

Then the operator S_ψ *is bounded on* $L^p(\mathbb{R}^n)$ *for all* $p \in (1,\infty)$.

If we put $Q(x) = [(\partial/\partial t)P_t(x)]_{t=1}$ for the Poisson kernel $P_t(x) = c_n t/(|x|^2 + t^2)^{(n+1)/2}$ of the upper half space $\mathbb{R}^n \times (0,\infty)$, then $S_Q(f)$ is the well-known Littlewood-Paley function $g(f)$ on $\mathbb{R}^n$. Also, for the Haar function $H(x) = \chi_{[-1,0]}(x) - \chi_{[0,1]}(x)$ on $\mathbb{R}$, $S_H(f)$ coincides with the Marcinkiewicz integral

$$\mu(f)(x) = \left(\int_0^\infty |F(x+t)+F(x-t)-2F(x)|^2 \frac{dt}{t^3}\right)^{1/2},$$

where $F(x) = \int_0^x f(y)\,dy$. We can easily see that the functions Q and H satisfy the conditions (5.3) and (5.4), and hence we can apply Theorem 13 to g and μ to get their L^p boundedness ($1 < p < \infty$).

It is known that condition (5.4) is superfluous. Indeed, if we put

$$B_\epsilon(\psi) = \int_{|x|\ge 1} |\psi(x)|\,|x|^{\epsilon}\,dx, \qquad \epsilon > 0,$$

$$D_u(\psi) = \left(\int_{|x|\le 1} |\psi(x)|^u\,dx\right)^{1/u}, \qquad u > 1,$$

we have the following (see [**21**], [**42**]):

THEOREM 14. *For* $\psi \in L^1(\mathbb{R}^n)$, *we assume the following*:
(1) $B_\epsilon(\psi) < \infty$ *for some* $\epsilon > 0$;
(2) $D_u(\psi) < \infty$ *for some* $u > 1$;
(3) *there exist non-negative functions* $h(r)$ *and* $\Omega(\theta)$ *such that* $|\psi(x)| \le h(r)\Omega(\theta)$ ($r = |x|, \theta = x/|x|$) *for all* $x \in \mathbb{R}^n \setminus \{0\}$, *where*
(4) $h(r)$ *is decreasing on* $(0,\infty)$;
(5) $h(|x|) \in L^1(\mathbb{R}^n)$;
(6) $\Omega \in L^q(S^{n-1})$ *for some* q *with* $1 < q \le \infty$.
Then S_ψ *is bounded on* $L^p(\mathbb{R}^n)$ *for all* $p \in (1,\infty)$.

THEOREM 15. *We assume that a function* ψ *satisfies the conditions* (1), (2) *of Theorem* 14. *We further assume that there exist non-negative functions* $h(r)$, $\Omega(\theta)$ *satisfying* (4)–(6) *of Theorem* 14 *for which we have* $|\psi(x)| \le h(r)\Omega(\theta)$ *for all* $|x| \ge 1$. *Then* S_ψ *is bounded on* $L^p(\mathbb{R}^n)$ *for all* $p \in [2,\infty)$. *In particular, if* ψ *is supported in* $\{|x| \le 1\}$, *then condition* (2) *of Theorem* 14 *alone implies the* L^p *boundedness of* S_ψ *for all* $p \in [2,\infty)$.

Consequently, by Theorem 14, the conditions (5.2), (5.3) only are sufficient for S_ψ to be bounded on $L^p(\mathbb{R}^n)$ for all $p \in (1,\infty)$. The case of the L^2 boundedness can be found in Coifman and Meyer [**11**]. In fact, it is known that if ψ satisfies these conditions, then S_ψ is also bounded on $L^p_w(\mathbb{R}^n)$ for all $p \in (1,\infty)$ and all $w \in A_p$, where A_p denotes the Muckenhoupt weight class (see [**42**]).

Theorems 14 and 15 follow from more general results. To see this, we now introduce two maximal functions:

$$N_\psi(f)(x) = \sup_{k\in\mathbb{Z}} |p_{2^k} * f(x)| \quad (x \in \mathbb{R}^n),$$
$$M_\psi(f)(x) = \sup_{t>0} ||\psi_t| * f(x)| \quad (x \in \mathbb{R}^n),$$

where $p(x) = \int_1^2 |\psi_t(x)|\, dt/t$. We note that $N_\psi(f) \le M_\psi(|f|)$. Also, if there exists an $\epsilon > 0$ such that

$$\left| \int_1^2 \mathcal{F}(|\psi|)(t\xi)\, dt \right| \le c|\xi|^{-\epsilon}, \tag{5.5}$$

$$\left| \int_1^2 \mathcal{F}(|\psi|)(t\xi)\, dt - \|\psi\|_1 \right| \le c|\xi|^{\epsilon}, \tag{5.6}$$

then Theorem A of [**16**] implies the L^p boundedness of N_ψ for all $p \in (1,\infty)$.

THEOREM 16. *If there exists an $\epsilon > 0$ such that*

$$\int_1^2 |\hat{\psi}(t\xi)|^2\, dt \le c\min\left(|\xi|^{\epsilon}, |\xi|^{-\epsilon}\right) \qquad (\xi \in \mathbb{R}^n \setminus \{0\}), \tag{5.7}$$

then S_ψ is bounded on $L^2(\mathbb{R}^n)$. If we further assume that N_ψ is bounded on $L^r(\mathbb{R}^n)$ for some $r \in (1,\infty)$, then S_ψ is bounded on $L^p(\mathbb{R}^n)$ for all $p \in (2, 2r/(r-1))$.

THEOREM 17. *Suppose that* (5.7) *holds and that M_ψ is bounded on $L^r(\mathbb{R}^n)$ for some $r \in (1,\infty)$. Then S_ψ is bounded on $L^p(\mathbb{R}^n)$ for all $p \in (2r/(r+1), 2)$.*

From condition (5.7), it follows that $\int_1^2 |\hat{\psi}(t\xi)|^2\, dt \le c|\xi|^{-\epsilon}$. These estimates are essentially less restrictive than the pointwise estimates. Indeed, we can find ψ such that

(1) $\int_1^2 |\hat{\psi}(t\xi)|^2\, dt \le c|\xi|^{-\epsilon}$ for some $\epsilon > 0$,

(2) there exists no $\delta > 0$ for which we have the pointwise estimate $|\hat{\psi}(\xi)| \le c|\xi|^{-\delta}$.

Also, there exist $r \in (1,\infty)$ and ψ such that N_ψ is bounded on L^r but M_ψ is not. By conditions (1), (2) of Theorem 14 and the condition $|\psi(x)| \le h(r)\Omega(\theta)$ ($|x| \ge 1$) of Theorem 15 we can show (5.7), (5.5), (5.6) (see [**42**]). Therefore, Theorem 15 follows from Theorem 16 (and the remark preceding it). In addition, if we assume that $|\psi(x)| \le h(r)\Omega(\theta)$ for all $x \in \mathbb{R}^n$, then M_ψ is bounded on $L^r(\mathbb{R}^n)$ for all $r \in (1,\infty)$, and so we can get Theorem 14 from Theorems 16 and 17.

Theorems 16 and 17 can be proved by extending the methods of Duoandikoetxea and Rubio de Francia [**16**] to the vector-valued setting (see [**21**]). Now we make some further remarks on L^2 boundedness. We can see that S_ψ is bounded on L^2 if and only if

$$\sup_{|\xi|=1} \int_0^\infty \left|\hat{\psi}(t\xi)\right|^2 dt/t < \infty,$$

but under certain suitable conditions on ψ, we have

$$\int_0^\infty \left|\hat{\psi}(t\xi)\right|^2 dt/t$$
$$= \iint_{\mathbb{R}^n\times\mathbb{R}^n} \psi(x)\overline{\psi(y)} \left(-\log|\langle\xi, x-y\rangle| - i\frac{\pi}{2}\operatorname{sgn}\langle\xi, x-y\rangle\right) dx\, dy.$$

By using this we obtain the following (see [**42**]).

PROPOSITION 2. *Suppose that* $\psi \in L^1(\mathbb{R}^n)$ *satisfies* (5.2). *Then if*

$$L(\psi) = \sup_{|\xi|=1} \iint_{\mathbb{R}^n \times \mathbb{R}^n} |\psi(x)\psi(y)|\, |\log|\langle \xi, x-y\rangle||\, dx\, dy < \infty,$$

S_ψ *is bounded on* $L^2(\mathbb{R}^n)$.

Let

$$\psi(x) = |x|^{-n+\rho}\Omega(x')\chi_{(0,1]}(|x|) \qquad (\rho > 0),$$

where $\Omega \in L^1(S^{n-1})$ is assumed to satisfy $\int \Omega(\theta)\, d\sigma(\theta) = 0$. If we put $\mu_\rho(f) = S_\psi(f)$, then $\mu_\rho(f)$ is known as the (generalized) Marcinkiewicz integral. Stein [**51**] introduced this (for $\rho = 1$) and proved some basic properties. Walsh [**58**] proved that μ_ρ is bounded on L^2 if $\int_{S^{n-1}} |\Omega|[\log(2+|\Omega|)]^{1/2}\, d\sigma < \infty$ (the author learned this fact from Ding [**14**]). This is an immediate consequence of the following result, which can be shown by using Proposition 2 along with Young's inequality.

PROPOSITION 3. *Suppose that* ψ *satisfies* (5.2) *and is compactly supported. If we further assume that*

$$\int_{\mathbb{R}^n} |\psi(x)|\, [\log(2+|\psi(x)|)]^{1/2}\, dx < \infty,$$

then S_ψ *is bounded on* $L^2(\mathbb{R}^n)$.

REMARK 3. Walsh [**58**] proved the L^2 boundedness of the Marcinkiewicz integral μ_ρ, assuming the condition

$$\sup_{\xi' \in S^{n-1}} \int_{S^{n-1}} |\Omega(\theta)| \left(\log \frac{1}{|\langle \xi', \theta\rangle|}\right)^{1/2} d\sigma(\theta) < \infty.$$

If $\int_{S^{n-1}} |\Omega|[\log(2+|\Omega|)]^{1/2}\, d\sigma < \infty$, this condition will be satisfied. In fact, we can find a more precise result in [**58**].

As for the weak $(1,1)$ boundedness of μ_ρ, a Marcinkiewicz integral analog of Theorem 3 is known:

THEOREM 18. *Suppose* $\Omega \in L\log L(S^{n-1})$. *Then* μ_ρ *is of weak type* $(1,1)$.

This is due to [**20**]. The proof was done by adapting the methods of Seeger [**48**] for the case of the Littlewood-Paley functions. We now give a generalization of Theorem 18. Suppose that ψ has support in $\{1 \le |x| \le 2\}$ and satisfies (5.2). For $t \in (0,1]$ and $q > 0$ put

$$\tilde{\omega}_q(t) = \sup_{|s|<t/2} \left(\int_{S^{n-1}} \left[\int_0^\infty |\psi((r-s)\theta) - \psi(r\theta)|^2\, dr/r \right]^{q/2} d\sigma(\theta) \right)^{1/q}.$$

When $q = \infty$, we can define $\tilde{\omega}_\infty(t)$ by the standard modification. Put

$$V(\theta) = \left(\int_0^\infty |\psi(r\theta)|^2\, dr/r \right)^{1/2} \qquad (\theta \in S^{n-1}).$$

Then by using the methods of [**20**], we get the following.

THEOREM 19. *If $V \in L\log L(S^{n-1})$ and $\int_0^1 \tilde{\omega}_1(t)\,dt/t < \infty$, then S_ψ is of weak type* $(1,1)$.

This was mentioned in a recent paper by Fan and Sato. Suppose that Ψ satisfies the hypotheses of Theorem 19 for ψ. Put $\psi(x) = \sum_{k\in\mathbb{Z}} c_k \Psi_{2^k}(x)$, where $\{c_k\}$ is a sequence of non-negative numbers such that $\sum_k c_k < \infty$. Then, since

$$S_\psi(f) \le \sum_k c_k S_{\Psi_{2^k}}(f) = (\sum_k c_k) S_\Psi(f),$$

the conclusion of Theorem 19 holds for S_ψ. If we put $\Psi(x) = |x|^{-n+\rho}\Omega(x')\chi_{(1,2]}(|x|)$ $(\rho > 0)$ and take $c_k = 2^{k\rho}$ $(k < 0)$, $c_k = 0$ $(k \ge 0)$, then we have

$$|x|^{-n+\rho}\Omega(x')\chi_{(0,1]}(|x|) = \sum_{k\in\mathbb{Z}} c_k \Psi_{2^k}(x),$$

and hence we can get Theorem 18 by using Theorem 19.

REMARK 4. Some recent research on the multilinear Littlewood-Paley function

$$T(f_1, f_2, \dots, f_m)(x) = \int_0^\infty ((\varphi_1)_t * f_1)(x)((\varphi_2)_t * f_2)(x) \cdots ((\varphi_m)_t * f_m)(x)\,dt/t$$

is in [**46**], where the mapping property of T is studied under assumptions for φ_j similar to those in Theorems 14 and 15 for ψ.

To conclude this note, we consider the Littlewood-Paley function related to the Bochner-Riesz means (see (3.2)), which is defined as

$$\sigma_\delta(f)(x) = \left(\int_0^\infty \left|S_R^\delta(f)(x) - S_R^{\delta-1}(f)(x)\right|^2 dR/R\right)^{1/2},$$

and also the Littlewood-Paley function related to the spherical means. We see the relation between them and how it can be used. For $\beta > 0$ define

$$M_t^\beta(f)(x) = c_\beta t^{-n} \int_{|y|<t} (1 - t^{-2}|y|^2)^{\beta-1} f(x-y)\,dy \qquad (f \in \mathcal{S}(\mathbb{R}^n)),$$

where $c_\beta = \Gamma(\beta + n/2)/(\pi^{n/2}\Gamma(\beta))$. By taking the Fourier transform, we can embed these operators in an analytic family of operators in β so that

$$M_t^0(f)(x) = c\int_{S^{n-1}} f(x - ty)\,d\sigma(y).$$

When $d_\beta = \beta + n/2 - 1 > 0$, put

$$\begin{aligned}\nu_\beta(f)(x) &= \left(\int_0^\infty \left|(\partial/\partial t)\ M_t^\beta(f)(x)\right|^2 t\,dt\right)^{1/2}\\ &= 2|d_\beta|\left(\int_0^\infty |M_t^\beta(f)(x) - M_t^{\beta-1}(f)(x)|^2\,dt/t\right)^{1/2}.\end{aligned}$$

Then the following result due to Kaneko and Sunouchi [**27**] is known.

THEOREM 20. *If* $\delta = \beta + n/2 - 1 > 0$, *then* $\sigma_\delta(f)$ *and* $\nu_\beta(f)$ $(f \in \mathcal{S}(\mathbb{R}^n))$ *are pointwise equivalent; that is, there are two positive constants* A *and* B *such that for all* $x \in \mathbb{R}^n$,

$$\sigma_\delta(f)(x) \le A\,\nu_\beta(f)(x) \le B\,\sigma_\delta(f)(x).$$

We now recall a result for σ_δ:

THEOREM 21. *If* $\delta > 1/2$ *and* $0 \le \alpha < 1$, *then*

$$\int_{\mathbb{R}^n} |\sigma_\delta(f)(x)|^2 |x|^{-\alpha}\,dx \le C_{\delta,\alpha} \int_{\mathbb{R}^n} |f(x)|^2 |x|^{-\alpha}\,dx.$$

This was shown by Carbery, Rubio de Francia, and Vega [**5**], and Rubio de Francia [**36**] gave a different proof. By Theorems 21 and 20 we can immediately get the following.

PROPOSITION 4. *If* $\beta > 3/2 - n/2$ *and* $0 \le \alpha < 1$, *then*

$$\int_{\mathbb{R}^n} |\nu_\beta(f)(x)|^2 |x|^{-\alpha}\,dx \le C_{\beta,\alpha} \int_{\mathbb{R}^n} |f(x)|^2 |x|^{-\alpha}\,dx.$$

Here we give an application of Proposition 4 to the maximal function

$$\mathcal{M}(f)(x) = \sup_{t>0} \left| \int_{S^{n-1}} f(x - ty)\,d\sigma(y) \right|.$$

We note that $\mathcal{M}(f)(x) = cM_*^0(f)(x)$, where $M_*^\beta(f)(x) = \sup_{t>0} \left| M_t^\beta(f)(x) \right|$. Duoandikoetxea and Vega [**17**] proved the following result.

THEOREM 22. *Suppose that* $n \ge 2$ *and* $n/(n-1) < p$. *Then the inequality*

$$\int_{\mathbb{R}^n} |\mathcal{M}(f)(x)|^p |x|^{-\alpha}\,dx \le C \int_{\mathbb{R}^n} |f(x)|^p |x|^{-\alpha}\,dx$$

holds for $n - p(n-1) < \alpha < n - 1$ *and does not hold for* $\alpha > n - 1$.

The affirmative part was partly proved by Rubio de Francia [**35**]. We note that the inequality for $\alpha = 0$ and $p > n/(n-1)$ was proved by Stein [**53**] for $n \ge 3$ and by Bourgain [**2**] for $n = 2$ (see also Sogge [**50**]). It is known that Proposition 4 can be used to prove the following.

PROPOSITION 5. *Suppose that* $\beta > 3/2 - n/2$ *and* $0 \le \alpha < 1$. *Then*

$$\int_{\mathbb{R}^n} \left| M_*^{\beta - 1/2}(f)(x) \right|^2 |x|^{-\alpha}\,dx \le C_{\beta,\alpha} \int_{\mathbb{R}^n} |f(x)|^2 |x|^{-\alpha}\,dx.$$

This is shown in [**53**] for $\alpha = 0$. It is noted in [**45**] that a different proof of the inequality in Theorem 22 for $n \ge 3$, $0 \le \alpha < n - 1$ and $p > n/(n-1)$ can be given by using Proposition 5 and an interpolation.

References

1. A. Benedek, A. P. Calderón and R. Panzone, *Convolution operators on Banach space valued functions*, Proc. Nat. Acad. Sci. U.S.A. **48** (1962), 356–365.
2. J. Bourgain, *Averages in the plane over convex curves and maximal operators*, J. Analyse Math. **47** (1986), 69–85.
3. A. P. Calderón and A. Zygmund, *On the existence of certain singular integrals*, Acta Math. **88** (1952), 85–139.
4. A. P. Calderón and A. Zygmund, *On singular integrals*, Amer. J. Math. **78** (1956), 289–309.
5. A. Carbery, J. L. Rubio de Francia and L. Vega, *Almost everywhere summability of Fourier integrals*, J. London Math. Soc. (2) **38** (1988), 513–524.
6. S. Chanillo and M. Christ, *Weak* $(1,1)$ *bounds for oscillatory singular integrals*, Duke Math. J. **55** (1987), 141–155.
7. L. K. Chen, *On a singular integral*, Studia Math. **85** (1987), 61–72.
8. M. Christ, *Weak type* $(1,1)$ *bounds for rough operators*, Ann. of Math. **128** (1988), 19–42.
9. M. Christ and J. L. Rubio de Francia, *Weak type* $(1,1)$ *bounds for rough operators, II*, Invent. Math. **93** (1988), 225–237.
10. M. Christ and C. D. Sogge, *The weak type* L^1 *convergence of eigenfunction expansions for pseudodifferential operators*, Invent. Math. **94** (1988), 421-453.
11. R. R. Coifman and Y. Meyer, *Au delà des opérateurs pseudo-différentiels*, Astérisque no. 57, Soc. Math. France, 1978.
12. R. R. Coifman and G. Weiss, *Extensions of Hardy spaces and their use in analysis*, Bull. Amer. Math. Soc. **83** (1977), 569–645.
13. W. Connett, *Singular integrals near* L^1, Proc. Symp. Pure Math. **35, Part 1** (1979), 163–165.
14. Y. Ding, *On Marcinkiewicz integral*, Proc. of the conference "Singular integrals and related topics, III" (Osaka Kyoiku University, Japan, January 27–29, 2001), 28–38.
15. Y. Ding, D. Fan and Y. Pan, *Weighted boundedness for a class of rough Marcinkiewicz integrals*, Indiana Univ. Math. J. **48** (1999), 1037–1055.
16. J. Duoandikoetxea and J. L. Rubio de Francia, *Maximal and singular integral operators via Fourier transform estimates*, Invent. Math. **84** (1986), 541–561.
17. J. Duoandikoetxea and L. Vega, *Spherical means and weighted inequalities*, J. London Math. Soc. (2) **53** (1996), 343–353.
18. D. Fan and Y. Pan, *Singular integral operators with rough kernels supported by subvarieties*, Amer. J. Math. **119** (1997), 799–839.
19. D. Fan and Y. Pan, *A singular integral operator with rough kernel*, Proc. Amer. Math. Soc. **125** (1997), 3695–3703.
20. D. Fan and S. Sato, *Weak type* $(1,1)$ *estimates for Marcinkiewicz integrals with rough kernels*, Tohoku Math. J. **53** (2001), 265–284.
21. D. Fan and S. Sato, *Remarks on Littlewood-Paley functions and singular integrals*, J. Math. Soc. Japan **54** (2002), 565–585.
22. R. Fefferman, *A note on singular integrals*, Proc. Amer. Math. Soc. **74** (1979), 266–270.
23. L. Grafakos and A. Stefanov, L^p *bounds for singular integrals and maximal singular integrals with rough kernels*, Indiana Univ. Math. J. **47** (1998), 455–469.
24. S. Hofmann, *Weak* $(1,1)$ *boundedness of singular integrals with nonsmooth kernel*, Proc. Amer. Math. Soc. **103** (1988), 260–264.
25. S. Hofmann, *Weighted weak-type* $(1,1)$ *inequalities for rough operators*, Proc. Amer. Math. Soc. **107** (1989), 423–435.
26. L. Hörmander, *Estimates for translation invariant operators in* L^p *spaces*, Acta Math. **104** (1960), 93–139.
27. M. Kaneko and G. Sunouchi, *On the Littlewood-Paley and Marcinkiewicz functions in higher dimensions*, Tôhoku Math. J. **37** (1985), 343–365.
28. J. E. Littlewood and R. E. A. C. Paley, *Theorems on Fourier series and power series*, J. London Math. Soc. **6** (1931), 230–233.
29. J. E. Littlewood and R. E. A. C. Paley, *Theorems on Fourier series and power series (II)*, Proc. London Math. Soc. **42** (1936), 52–89.
30. J. E. Littlewood and R. E. A. C. Paley, *Theorems on Fourier series and power series (III)*, Proc. London Math. Soc. **43** (1937), 105–126.

31. S. Z. Lu and Y. Zhang, *Criterion on L^p-boundedness for a class of oscillatory singular integrals with rough kernels*, Rev. Mat. Iberoamericana **8** (1992), 201–219.
32. J. Namazi, *On a singular integral*, Proc. Amer. Math. Soc. **96** (1986), 421–424.
33. F. Ricci and E. M. Stein, *Harmonic analysis on nilpotent groups and singular integrals, I*, J. Func. Anal. **73** (1987), 179–194.
34. F. Ricci and G. Weiss, *A characterization of $H^1(\Sigma_{n-1})$*, Proc. Symp. Pure Math. **35, Part 1** (1979), 289–294.
35. J. L. Rubio de Francia, *Weighted norm inequalities for homogeneous families of operators*, Trans. Amer. Math. Soc. **275** (1983), 781–790.
36. J. L. Rubio de Francia, *Transference principles for radial multipliers*, Duke Math. J. **58** (1989), 1–19.
37. J. L. Rubio de Francia, F. J. Ruiz and J. L. Torrea, *Calderón-Zygmund theory for operator-valued kernels*, Adv. Math. **62** (1986), 7–48.
38. S. Sato, *Spherical summability and a vector-valued inequality*, Bull. London Math. Soc. **27** (1995), 58–64.
39. S. Sato, *A weighted vector-valued weak type* $(1,1)$ *inequality and spherical summation*, Studia Math. **109** (1994), 159-170.
40. S. Sato, *Divergence of the Bochner-Riesz means in the weighted Hardy spaces*, Studia Math. **118** (1996), 261–275.
41. S. Sato, *Some weighted weak type estimates for rough operators*, Math. Nachr. **187** (1997), 211–240.
42. S. Sato, *Remarks on square functions in the Littlewood-Paley theory*, Bull. Austral. Math. Soc. **58** (1998), 199–211.
43. S. Sato, *Weighted weak type* $(1,1)$ *estimates for oscillatory singular integrals*, Studia Math. **141** (2000), 1–24.
44. S. Sato, *Weighted weak type* $(1,1)$ *estimates for oscillatory singular integrals with Dini kernels*, Bull. Fac. Ed. Kanazawa Univ. Natur. Sci. **49** (2000), 1–22.
45. S. Sato, *Some weighted estimates for Littlewood-Paley functions and radial multipliers*, J. Math. Anal. Appl. **278** (2003), 308-323.
46. S. Sato and K. Yabuta, *Multilinearized Littlewood-Paley operators*, Sci. Math. Jpn. **55** (2002), 447–453.
47. A. Seeger, *Endpoint estimates for multiplier transformations on compact manifold*, Indiana Univ. Math. J. **40** (1991), 471–533.
48. A. Seeger, *Singular integral operators with rough convolution kernels*, J. Amer. Math. Soc. **9** (1996), 95–105.
49. A. Seeger and T. Tao, *Sharp Lorentz space estimates for rough operators*, Math. Ann. **320** (2001), 381-415.
50. C. D. Sogge, *Fourier integrals in classical analysis*, Cambridge University Press, 1993.
51. E. M. Stein, *On the functions of Littlewood-Paley, Lusin, and Marcinkiewicz*, Trans. Amer. Math. Soc. **88** (1958), 430–466.
52. E. M. Stein, *Singular Integrals and Differentiability Properties of Functions*, Princeton Univ. Press, 1970.
53. E. M. Stein, *Maximal functions: Spherical means*, Proc. Nat. Acad. Sci. U.S.A. **73** (1976), 2174–2175.
54. E. M. Stein, *Harmonic Analysis: Real-Variable Methods, Orthogonality, and Oscillatory Integrals*, Princeton Univ. Press, 1993.
55. Q. Sun, *Two problems about singular integral operators*, Approx. Theory Appl. 7 (1991), 83–98.
56. T. Tao, *The weak-type* $(1,1)$ *of* $L\log L$ *homogeneous convolution operator*, Indiana Univ. Math. J. **48** (1999), 1547–1584.
57. A. Vargas, *Weighted weak type* $(1,1)$ *bounds for rough operators*, J. London Math. Soc. (2) **54** (1996), 297–310.

58. T. Walsh, *On the function of Marcinkiewicz. Collection of articles honoring the completion by Antoni Zygmund of 50 years of scientific activity, III*, Studia Math. **44** (1972), 203–217.
59. A. Zygmund, *Trigonometric series,* 2nd ed., Cambridge Univ. Press, Cambridge, London, New York and Melbourne, 1977.

DEPARTMENT OF MATHEMATICS, FACULTY OF EDUCATION, KANAZAWA UNIVERSITY, KANAZAWA 920-1192, JAPAN
E-mail address: shuichi@kenroku.kanazawa-u.ac.jp

Translated by SHUICHI SATO

Amer. Math. Soc. Transl.
(2) Vol. **215**, 2005

Convergence of metric measure spaces and energy forms

Atsushi Kasue

Introduction

Let us consider the Dirichlet problem for harmonic functions in a relatively compact domain of a Riemannian manifold. The solution is obtained by the direct method of the calculus of variations as follows. We first take a sequence of functions with prescribed boundary values which minimizes the Dirichlet energy. Then it follows from the Poincaré inequality that the sequence is bounded in the Sobolev space, and hence, by Rellich's theorem, we have a subsequence which converges to a function weakly in the Sobolev space and strongly in L^2. The lower semicontinuity of the Dirichlet energy implies that the limit function minimizes the energy and satisfies the Laplace equation in a weak sense. Second, we can obtain an *a priori* eatimate on the Hölder continuity of the solution with respect to the Riemannian distance, using Moser's iteration method, in which the volume estimates of metric balls, the Sobolev inequality, and Poincaré inequality play crucial roles.

Now Riemannian manifolds are metric spaces equipped with their Riemannian distances, and also Dirichlet spaces with the Dirichler energy functionals on the Sobolev spaces. From the former point of view, the convergence of Riemannian manifolds with respect to the Gromov-Hausdorff distance has been intensively studied. There are survey articles, for instance, [**16**], [**48**], [**49**] on the collapsing phenomena, [**39**] on the convergence of Einstein manifolds, and [**42**] on Riemannian manifolds with Ricci curvature bounded below. Also we refer to the monographs [**6**], [**18**], [**19**] for discussions in detail.

On the other hand, Fukaya [**15**] studied a family of compact Riemannian manifolds with sectional curvature bounded uniformly in absolute value, and showed the continuity of eigenvalues and also eigenspaces in a certain sense with respect to the topology of the measured Gromov-Hausdorff convergence. In a series of papers [**9**], J. Cheeger and T.H. Colding proved the continuity of the Dirichlet energy functionals in a certain sense with respect to the same topology under the assumption that the Ricci curvature of compact Riemannian manifolds in a given family is uniformly bounded below.

This article originally appeared in Japanese in Sūgaku **55** (1) (2003), 20–36.
2000 *Mathematics Subject Classification.* Primary 58Jxx; Secondary 53Cxx.

From the viewpoints of the direct method of calculus of variations mentioned above and also the convergence of the Gromov-Hausdorff sense, we would like to discuss the convergence of the Dirichlet energy functionals on Riemannian manifolds and more generally certain Dirichlet spaces. In Section 1, we recall some basic facts on the Gromov-Hausdorff distance and state a theorem due to Cheeger [**8**] on the differentiability of Lipschitz functions on metric measure spaces satisfying the so-called volume doubling property and the (scale invariant) Poincaré inequality. Section 2 is devoted to the introduction of regular Dirichlet spaces with the intrinsic distances. In Section 3, the convergence of closed forms on Hilbert spaces in the sense of Mosco and their spectral structures are discussed, and some results concerning precompactness of a family of Dirichlet spaces are given. We also study the convergence of Riemannian vector bundles and the energy functionals on the space of maps in Section 4.

1. Measured Gromov-Hausdorff convergence

To begin with, we recall the definition of Gromov-Hausdorff distance between metric spaces. Given two metric spaces (X, d_X) and (Y, d_Y), a (Borel measurable) map $f : X \to Y$ of X to Y is called an ε-Hausdorff approximation for a positive number ε if it satisfies $|d_X(x, x') - d_Y(f(x), f(x'))| \leq \varepsilon$, $x, x' \in X$, and $Y \subset f(X)_\varepsilon$, where $f(X)_\varepsilon$ stands for the ε-neighborhood of the subset $f(X)$ in Y. We denote by $HD(X, Y)$ the infimum of the positive numbers ε for which there exists an ε-Hausdorff approximation of X to Y. We observe that $HD(X, \overline{X}) = 0$ for a bounded metric space X and its completion $\overline{X}$, and also that for compact metric spaces X and Y, $HD(X, Y) = 0$ if and only if they are isometric.

Let $X = (X, d_X)$ be a metric space. A subset S of X is said to be ε-separated for a positive constant ε if $d_X(a, b) > \varepsilon$ for all $a, b \in S$, $a \neq b$. Given a continuous curve $c : [a, b] \to X$, we denote by $L(c)$ the least upper bound for $\sum_{i=0}^{n-1} d_X(c(t_i), c(t_{i+1}))$, where $a = t_0 < t_1 < \cdots < t_n = b$, and when $L(c)$ is finite, we say that c is rectifiable and of length $L(c)$. We call X a length space if for any pair of points $p, q \in X$, the distance $d_X(p, q)$ is equal to the greatest lower bound of the lengths of all rectifiable curves joining p to q. In the case where X is a locally compact length space, the Hopf-Rinow theorem asserts that the following are equivalent: (1) X is complete; (2) any bounded closed subset of X is compact; (3) any locally distance-minimizing curve $c : [0, a) \to X$ can be continuously extended to the closed interval $[0, a]$; (4) there exists a point such that every distance-minimizing curve $c : [0, a) \to X$ emanating from p can be continuously extended to the closed interval $[0, a]$. Moreover, if one of these properties holds, then any pair of points of X can be joined by distance-minimizing curves.

The following are basic results by M. Gromov.

THEOREM 1.1. (1) *Let $\mathcal{F}$ be a family of compact metric spaces. Suppose that $\mathcal{F}$ is uniformly and totally bounded; that is, the diameters of all X in $\mathcal{F}$ are bounded by a constant and for any positive number ε, there exists a positive constant $N(\varepsilon)$ such that the cardinality of an ε-separated subset of X is bounded by $N(\varepsilon)$. Then $\mathcal{F}$ is precompact with respect to the Gromov-Hausdorff distance.*

(2) *Let $\{X_n\}$ be a sequence of metric spaces which converges to a complete metric space X with respect to the Gromov-Hausdorff distance. If each X_n is a length space, then so is X.*

Let us outline the proof of the first assertion. Let $\{X_n\}$ be a sequence in $\mathcal{F}$. Given a positive number ε, we take a maximal ε-separated subset from each X_n. Then the sequence of such subspaces is precompact because of the assumptions, so that a subsequence of such subspaces converges to a finite metric space. Taking a sequence of positive numbers $\{\varepsilon_m\}$ tending to zero and using a diagonal argument, we can obtain a limit metric space of the subspaces whose completion belongs to the boundary of the sequence $\{X_n\}$.

This argument suggests that any compact length space is a limit of finite metric graphs with respect to the Gromov-Hausdorff distance. See [**6**], [**18**], [**19**] for details and further results on this subject.

Now we are able to define the Gromov-Hausdorff distance between two metric spaces if the diameters are bounded. Otherwise, it is not clear whether the distance can be determined. For this reason, let us introduce a topology on a set of pointed metric spaces as follows. We say a sequence $\{X_n = (X_n, o_n, d_n)\}$ of pointed metric spaces converges to such a space $X = (X, o, d)$ in the Gromov-Hausdorff sense if there exist sequences of positive numbers $\{R_n\}$ with $\lim_{n\to\infty} R_n = +\infty$ and $\{\varepsilon_n\}$ with $\lim_{n\to\infty} \varepsilon_n = 0$, and a sequence of maps $f_n : B(o_n, R_n) \to X$ such that (1) $f_n(o_n) = o$, (2) $|d_n(x,y) - d(f_n(x), f_n(y))| < \varepsilon_n$, $x, y \in B(o_n, R_n)$, and (3) $B(o, R_n - \varepsilon_n) \subset f_n(B(o_n, R_n))_{\varepsilon_n}$. In addition, these approximating maps allow us to define the convergence of functions and subsets as follows. Given a bounded subset U in X and a function u on U, we say that a sequence of bounded subsets U_n of X_n converges to U if f_n gives rise to an ε_n-Hausdorff approximation between U_n and U, and a sequence of functions u_n defined on a subset including $f_n^{-1}(U)$ in X_n uniformly converges to u if $\sup\{|u_n(x) - u(f_n(x))| \mid x \in f_n^{-1}(U)\}$ tends to zero as $n \to \infty$.

In what follows, we consider locally compact, separable, and complete metric spaces (X, d) and Radon measures μ which satisfy $0 < \mu(B(x,r)) < +\infty$, $x \in X, r > 0$. We say that a sequence $\{X_n = (X_n, o_n, d_n, \mu_n)\}$ of pointed metric spaces endowed with Radon measures μ_n converges to such a space $X = (X, o, d, \mu)$ in the measured Gromov-Hausdorff sense if (X_n, o_n) converges to (X, o) in the Gromov-Hausdorff sense and the push-forward of the measures μ_n by approximating maps $f_n : X_n \to X$ vaguely converges to μ as $n \to \infty$.

Now, before recalling some fundamental results in [**8**], we consider two properties of a metric measure space (X, d, μ). First, we say that (X, d, μ) has the volume doubling property if, for some positive constants κ and R,

$$[D]_\kappa \qquad \mu(B(x,2r)) \le 2^\kappa \mu(B(x,r)), \quad x \in X,\ 0 < r < R.$$

Second, for a locally Lipschitz continuous function u, putting

$$\mathrm{dil}_d\, u\,(x) = \limsup_{r\to 0}\{|u(y) - u(z)|/d(y,z) \mid y, z \in B(x,r),\ y \ne z\},$$

the L^p integral of $\mathrm{dil}_d u$ on a bounded measurable subset A, where $p \in [1, +\infty]$, can be defined and called the p-Dirichlet energy of u over A. Then we say that (X, d, μ) satisfies a weak p-Poincaré inequality if, for some positive constants τ and R and for any locally Lipschitz continuous function u,

$[P]_\tau$

$$\left(\fint_{B(x,r)} |u - u_{x,r}|^p \, d\mu\right)^{1/p} \le \tau r \left(\fint_{B(x,2r)} (\mathrm{dil}_d u)^p \, d\mu\right)^{1/p}, \quad x \in X,\ 0 < r < R,$$

where we put $⨍_A u\,d\mu = \mu(A)^{-1}\int_A u\,d\mu$ and $u_{x,r} = ⨍_{B(x,r)} u\,d\mu$.

The following results are proved in Cheeger [**8**].

THEOREM 1.2. *Let $X_n = (X_n, o_n, d_n, \mu_n)$ be a sequence of pointed measured metric spaces which converges to a pointed metric measure space (X, o, d, μ). If all X_n satisfy the volume doubling property $[\mathrm{D}]_\kappa$, then so does X, and further if the weak p-Poincaré inequality $[\mathrm{P}]_\tau$ holds on each X_n, it also holds on X.*

This theorem is an immediate consequence of the following.

LEMMA 1.3. *Let $X_n = (X_n, o_n, d_n, \mu_n)$ and $X = (X, o, d, \mu)$ be as in Theorem 1.2. Suppose that each X_n satisfies the volume doubling property $[\mathrm{D}]_\kappa$. Let u be a locally Lipschitz continuous function defined on a bounded open subset U in X. Then there exists a sequence of locally Lipschitz continuous functions u_n on bounded open subsets U_n in X_n such that u_n uniformly converges to u as $n \to \infty$ and moreover*

$$\int_U (\mathrm{dil}_d u)^p\,d\mu = \lim_{n\to\infty} \int_{U_n} (\mathrm{dil}_{d_n} u_n)^p\,d\mu_n.$$

We remark that in the situation of this lemma, for a sequence of locally Lipschitz continuous functions $\xi_n : U_n \to \mathbf{R}$ which uniformly converges to such a function $\xi : U \to \mathbf{R}$, the lower semicontinuity of the p-Dirichlet energy functionals does not hold true; that is, *the following does not hold in general*:

$$\int_U (\mathrm{dil}_d\,\xi)^p\,d\mu \le \liminf_{n\to\infty} \int_{U_n} (\mathrm{dil}_{d_n}\,\xi_n)^p\,d\mu_n.$$

Now Rademacher's theorem on the differentiation of Lipschitz functions can be expanded as follows.

THEOREM 1.4. *Let (X, d, μ) be a metric measure space with the volume doubling property $[\mathrm{D}]_\kappa$ and the weak p-Poincaré inequality $[\mathrm{P}]_\tau$ for some constants κ, τ and $p \in (1, +\infty)$. Then X can be covered by a union of Borel subsets U_i ($i = 1, 2, \ldots$) and Z with $\mu(Z) = 0$ in such a way that for each i, there exists a linearly independent set $\{u_1^{(i)}, \ldots, u_{n(i)}^{(i)}\}$ of Lipschitz continuous functions on U_i with the following properties:*

(1) The number $n(i)$ of the functions is bounded above by a constant $N(\kappa, \tau, p)$ depending only on the given ones.

(2) For any $(a) = (a_1, \ldots, a_{n(i)}) \in \mathbf{R}^{n(i)}$ with $(a) \neq 0$, the Lipschitz function $u_{(a)} = \sum_{j=1}^{n(i)} a_j u_j^{(i)}$ is asymptotically generalized linear at each point $x \in U_i$ and $\mathrm{dil}_d\,u\,(x) > 0$; moreover for any locally Lipschitz function u, there exists an $\mathbf{R}^{n(i)}$-valued measurable function $a(u; x)$ ($x \in U_i$) such that

$$\lim_{y\to x} \frac{u(y) - u(x) - \sum_{j=1}^{n(i)} a_j(u, x)(u_j^{(i)}(y) - u_j^{(i)}(x))}{d(y, x)} = 0.$$

In this theorem, we say that a locally Lipschitz function u is asymptotically generalized linear at a point x if it belongs to the Lebesgue points of $(\mathrm{dil}_d\,u)^p$ and

$$\lim_{r\to 0}\left(⨍_{B(x,r)} (\mathrm{dil}_d u)^p\,d\mu - \inf_{k\in C_0^{0,1}(B(x,r))} ⨍_{B(x,r)} (\mathrm{dil}_d(u+k))^p\,d\mu \right) = 0.$$

Let u be a locally Lipschitz continuous function on X. According to Theorem 1.4, the function $x \in U_i \to a(u;x) = (a_j(u;x)) \in \mathbf{R}^{n(i)}$ can be thought of as a section of the cotangent bundle T^*X of X on U_i, and we obtain a section of T^*X defined almost everywhere on X, which will be denoted by du. Then we see that for $u, v \in C^{0,1}_{loc}(X)$ and $\alpha, \beta \in \mathbf{R}$, $d(\alpha u + \beta v) = \alpha du + \beta dv$ and $d(uv) = vdu + udv$. In addition, $|du| = \mathrm{dil}_d\, u$ defines a canonical norm on T^*X to which the Dirichlet 2-energy $\int_X (\mathrm{dil}_d\, u)^2 d\mu$ is associated. This norm is not induced from a quadratic form on T^*X in general; however, we can define an L^∞ Riemannian metric $\langle\ ,\ \rangle$ on T^*X in such a way that for some positive constants λ and Λ depending only on the given κ and τ, $\lambda\ \mathrm{dil}_d\, u \le \sqrt{\langle du, du\rangle} \le \Lambda\ \mathrm{dil}_d\, u$ for any $u \in C^{0,1}_{loc}(X)$. Then on the space of locally Lipschitz continuous functions of bounded Dirichlet 2-energy, we have a quadratic form

$$\mathcal{E}(u,v) = \int_X \langle du, dv\rangle\, d\mu,$$

which can be extended to a densely defined closed form on the Hilbert space $L^2(X,\mu)$. We denote the form by the same letter and the domain by $D[\mathcal{E}]$. Then the form $(\mathcal{E}, D[\mathcal{E}])$ in fact gives a strongly local, regular Dirichlet form on $L^2(X,\mu)$, and $C^{0,1}_{loc}(X) \cap D[\mathcal{E}]$ turns out to be a core of the form.

We recall here some basic notions on Dirichlet spaces. Let X be a locally compact, separable metric space and μ a Radon measure on X with $\mathrm{supp}\,\mu = X$. Given a densely defined closed form $\mathcal{E}$ on $L^2(X,\mu)$, we call it a Dirichlet form if for any u in the domain $D[\mathcal{E}]$ of $\mathcal{E}$, $v = \max\{\min\{u,1\},0\}$ also belongs to $D[\mathcal{E}]$ and $\mathcal{E}(v,v) \le \mathcal{E}(u,u)$. A subspace $\mathcal{C}$ of $C_0(X) \cap D[\mathcal{E}]$ is called a core of $\mathcal{E}$ if it is dense in $D[\mathcal{E}]$ with respect to the norm $(\mathcal{E}(\cdot,\cdot) + \|\cdot\|^2_{L^2})^{1/2}$ and also dense in $C_0(X)$ equipped with the L^∞ norm. When $\mathcal{E}$ admits a core, we say $\mathcal{E}$ is regular. If $\mathcal{E}(u,v) = 0$ for any $u, v \in D[\mathcal{E}]$ such that $\mathrm{supp}\, u \cap \mathrm{supp}\, v = \emptyset$ (resp., u is constant on $\mathrm{supp}\, v$), then the form is said to be local (resp. strongly local).

We refer the reader to [**17**], Part I, for basic properties of Dirichlet spaces, and also to [**5**], [**20**], [**21**], [**22**], [**35**], etc., for many results and interesting examples concerning the volume doubling property and the weak Poincaré inequality. Finally, we should mention that in [**8**], in addition to the extension of Rademarcher's theorem as above, the differentiation of Lipschitz continuous functions is expressed in terms of tangent cones.

2. Dirichlet forms and intrinsic distances

In this section, we consider a regular Dirichlet space $(X, \mu, \mathcal{E})$. For any $u \in D[\mathcal{E}] \cap L^\infty$, there exists a unique positive Radon measure $\mu_{\langle u,u\rangle}$ given by

$$\phi \in D[\mathcal{E}] \cap C_0(X) \longrightarrow \mathcal{E}(u, \phi u) - \frac{1}{2}\mathcal{E}(u^2, \phi).$$

We call it the energy measure of u. If we set $\mu_{\langle u,v\rangle} = \frac{1}{2}(\mu_{\langle u+v,u+v\rangle} - \mu_{\langle u,u\rangle} - \mu_{\langle v,v\rangle})$ for $u, v \in D[\mathcal{E}] \cap L^\infty$, then this defines a bounded signed measure on X.

Let us assume that $\mathcal{E}$ is strongly local. In this case, the energy measure of a function $u \in D[\mathcal{E}] \cap L^\infty$ vanishes in an open subset A if u is constant there. This implies that the energy measure restricted to A depends only on the restriction of the function there. This allows us to define a space $D_{loc}[\mathcal{E}]$ which consists of locally square summable functions such that for any open subset A, there exists $v \in D[\mathcal{E}]$ which coincides with u in A. For a function $u \in D_{loc}[\mathcal{E}]$, the energy measure $\mu_{\langle u,u\rangle}$

on X can be defined. Let us denote by $\mathcal{A}[\mathcal{E}]$ the space of functions in $D_{loc}[\mathcal{E}]$ such that the $\mu_{\langle u,u\rangle}$ are absolutely continuous with respect to the measure μ. Namely, we put

$$\mathcal{A}[\mathcal{E}] = \{u \in D_{loc}[\mathcal{E}] \mid \mu_{\langle u,u\rangle} = \Gamma(u,u)\mu,\ \Gamma(u,u) \in L^1_{loc}(X)\}.$$

Let us further introduce a pseudo-distance, called the intrinsic (pseudo-)distance, by

$$d_{\mathcal{E}}(x,y) = \sup\{u(x) - u(y) \mid u \in \mathcal{A}[\mathcal{E}] \cap C(X),\ \Gamma(u,u) \le 1 \quad \mu\text{-a.e.}\}$$

(cf. [**4**], [**10**]). The following is proved in Sturm [**45**].

THEOREM 2.1. *Let $(X,\mu,\mathcal{E})$ be a strongly local, regular Dirichlet space. Assume the property* [C] : *the intrinsic pseudo-distance $d_{\mathcal{E}}$ defines a distance on X and induces the same topology of X. Then the following assertions hold: (1) $(X, d_{\mathcal{E}})$ becomes a length space, (2) for a compact subset Y, the distance function to Y, $\rho_Y(x) = \inf\{d_{\mathcal{E}}(x,y) \mid y \in Y\}$ $(x \in X)$, belongs to $\mathcal{A}[\mathcal{E}] \cap C(X)$ and satisfies $\Gamma(\rho_Y, \rho_Y) \le 1$.*

We note that for the Riemannian distance $d_{\mathcal{E}}$ on a Riemannian manifold we have $\Gamma(\rho_Y,\rho_Y) = 1$; however, if we admit the coefficients of measurable functions, then the equality is not true in general (cf. [**47**]).

What the second assertion of the above theorem suggests is clarified in the following.

THEOREM 2.2. *Let $(X,\mu,\mathcal{E})$ be a strongly local, regular Dirichlet space satisfying property* [C] *and the volume doubling property* $[\mathrm{D}]_\kappa$ *for some constants $\kappa > 0$ and $R \in (0,+\infty]$ with respect to the intrinsic distance $d_{\mathcal{E}}$. Then $C^{0,1}_{loc}(X, d_{\mathcal{E}}) \subset \mathcal{A}[\mathcal{E}]$, and*

$$\Gamma(u,u)^{1/2} \le \mathrm{dil}_{d_{\mathcal{E}}} u \quad \mu\text{-a.e.}$$

for $u \in C^{0,1}_{loc}(X, d_{\mathcal{E}})$. Moreover, suppose that for some constants $\tau > 0$ and $R \in (0,+\infty]$, the following weak Poincaré inequality on the Dirichlet form $\mathcal{E}$ holds:

$$([P]_{\mathcal{E},\tau}) \qquad \left(\fint_{B(x,r)} |u - u_{x,r}|^2 \, d\mu \right)^{1/2} \le \tau r \left(\fint_{B(x,2r)} \mu_{\langle u,u\rangle} \right)^{1/2}.$$

Then there exists a positive constant C, depending only on κ and τ, such that

$$C \ \mathrm{dil}_{d_{\mathcal{E}}} u \le \Gamma(u,u)^{1/2} \quad \mu\text{-a.e.}$$

for $u \in C^{0,1}_{loc}(X, d_{\mathcal{E}})$.

Let $(X,\mu,\mathcal{E})$ be a strongly local, regular Dirichlet space with the properties in this theorem. Then we can apply Cheeger's result (Theorem 1.4) to X and obtain an L^∞ Riemannian metric on the cotangent bundle T^*X by putting $\langle du, du\rangle = \Gamma(u,u)$ $(u \in C^{0,1}_{loc}(X, d_{\mathcal{E}}))$. Now we have the following volume estimate under conditions $[\mathrm{D}]_\kappa$ and $[\mathrm{P}]_{\mathcal{E},\tau}$.

THEOREM 2.3. *Let $(X,\mu,\mathcal{E})$, κ, and τ be as in Theorem 2.2. Then there exists a positive constant C', depending only on κ and τ, such that*

$$\mu(B(x,D)) \le \mu(B(x,r)) \exp C'(1 + D/r), \quad x \in X,\ 0 < r < R,\ r < D.$$

According to Sturm [**46**] (see also [**41**] and the references therein), for a strongly local, regular Dirichlet space $(X, \mu, \mathcal{E})$ satisfying property [C], X satisfies properties $[\mathrm{D}]_\kappa$ and $[\mathrm{P}]_\tau$ if and only if a parabolic Harnack inequality holds on X. Making use of the inequality and the argument in Ishige and Murata [**24**], we are able to verify Theorem 2.3. For details on the theorem and also Theorem 2.2, we refer the reader to [**26**] and [**28**].

Now let us consider a complete Riemannian manifold of dimension k whose Ricci curvature is bounded from below by a nonpositive constant $-(k-1)K^2$ $(K \geq 0)$. Then the Bishop-Gromov volume comparison theorem reads

$$\frac{\mathrm{Vol}(B(x,D))}{\int_0^D (\sinh Kt)^{k-1} dt} \leq \frac{\mathrm{Vol}(B(x,r))}{\int_0^r (\sinh Kt)^{k-1} dt}, \qquad 0 < r \leq D < +\infty.$$

Furthermore, the isoperimetric inequality in Buser [**7**] shows that

$$\int_{B(x,r)} |u - u_{x,y}| \, dv_M \leq Cr \exp(Kr) \int_{B(x,r)} |\nabla u| \, dv_M$$

for all $u \in C^\infty(M)$, $x \in M$, and $0 < r < +\infty$. Therefore if we fix a positive constant R, then there exist positive constants κ and τ, depending only on k, R, and K, such that the volume doubling property $[\mathrm{D}]_\kappa$ and the weak Poincaré inequality $[\mathrm{P}]_\tau$ hold in any ball $B(x,R)$. In particular, in case $K = 0$, we may take $R = +\infty$; that is, $[\mathrm{D}]_\kappa$ and $[\mathrm{P}]_\tau$ hold globally. Examples of spaces possessing such properties other than complete Riemannian manifolds with nonnegative Ricci curvature are provided by limit spaces of such Riemannian manifolds (cf. [**37**]), Carnot groups, which also appear as tangent cones of sub-Riemannian manifolds (cf. [**2**]), and tangent cones of the spaces in Theorem 1.4 and Theorem 2.2.

3. Convergence of Dirichlet forms

In this section, following Kuwae and Shioya [**34**], we first give the definition of a convergence of quadratic forms of Hilbert spaces and describe some consequences of the convergence. After that we state some results on the convergence of Dirichlet spaces.

Let us consider a sequence of separable Hilbert spaces $\{H_n\}$, a separable Hilbert space H, and a dense subspace $\mathcal{C} \subset H$ such that there exist linear maps Φ_n of $\mathcal{C}$ to H_n satisfying

[*] $$\lim_{n\to\infty} |\Phi_n(u)|_{H_n} = |u|_H, \qquad u \in \mathcal{C}.$$

The existence of such linear maps allows us to define some notions on convergence of functions, functionals, and operators as follows:

DEFINITION 3.1. (1) A sequence of $u_n \in H_n$ strongly converges to $u \in H$ as $n \to \infty$ if there exists a sequence $\{\tilde{u}_k\}$ in $\mathcal{C}$ tending to u as $k \to \infty$ such that

$$\lim_{k\to\infty} \limsup_{n\to\infty} |\Phi_n(\tilde{u}_k) - u_n|_{H_n} = 0.$$

(2) A sequence of $u_n \in H_n$ weakly converges to $u \in H$ as $n \to \infty$ if

$$\lim_{n\to\infty} \langle u_n, v_n \rangle_{H_n} = \langle u, v \rangle_H$$

for every $v \in H$ and any sequence of $v_n \in H_n$ which strongly converges to v as $n \to \infty$.

(3) A sequence of bounded linear operators $B_n : H_n \to H_n$ strongly converges to a bounded linear operator $B : H \to H$ as $n \to \infty$ if $B_n(u_n)$ strongly converges to $B(u)$ for every $u \in H$ and any sequence of $u_n \in H_n$ which strongly converges to u.

(4) A sequence of quadratic forms $\mathcal{E}_n$ on H_n converges, as $n \to \infty$, to a quadratic form $\mathcal{E}$ on H in the sense of Mosco if the following two conditions hold:

(i) if a sequense of $u_n \in H_n$ weakly converges to $u \in H$, then

$$\mathcal{E}(u,u) \le \liminf_{n\to\infty} \mathcal{E}_n(u_n,u_n) \ (\le +\infty);$$

(ii) for any $u \in H$, there exists a sequence of $u_n \in H_n$ which strongly converges to u in such a way that

$$\limsup_{n\to\infty} \mathcal{E}_n(u_n,u_n) \le \mathcal{E}(u,u) \ (\le +\infty).$$

(5) A sequence of quadratic forms $\mathcal{E}_n$ on H_n is asymptotically compact if any sequence of $u_n \in H_n$ with $\sup_n |u_n|^2_{H_n} + \mathcal{E}_n(u_n,u_n) < +\infty$ contains a subsequence which strongly converges (to some $u \in H$).

In the case where the weak convergence of a sequence of elements $u_n \in H_n$ in condition (i) of Definition 3.1 (4) is replaced by the strong convergence of $\{u_n\}$, we say that the sequence of forms $\mathcal{E}_n$ Γ-converges to a form $\mathcal{E}$.

We recall that any sequence of quadratic forms on H_n admits a subsequence which Γ-converges to a closed form on H (cf., e.g., [**11**]).

We remark that if $\{\mathcal{E}_n\}$ is asymptotically compact and it Γ-converges to $\mathcal{E}$ as $n \to \infty$, then it converges to $\mathcal{E}$ in the sense of Mosco.

THEOREM 3.1. *Let $\mathcal{E}_n$ and $\mathcal{E}$ be densely defined closed forms on H_n and H, respectively. Let $G_{n;\zeta}$, $P_{n;t}$, $\mathcal{L}_n$, and E_n be respectively the resolvent, the semigroup, the infinitesimal generator, and the spectral resolution associated with $\mathcal{E}_n$, and also let G_ζ, P_t, $\mathcal{L}$, and E be respectively those of $\mathcal{E}$.*

(1) *The following conditions are equivalent:*

(i) *$\mathcal{E}_n$ converges to $\mathcal{E}$ in the sense of Mosco as $n \to \infty$.*

(ii) *For some $\zeta < 0$, $G_{n;\zeta}$ strongly converges to G_ζ as $n \to \infty$.*

(iii) *For some $t > 0$, $P_{n;t}$ strongly converges to P_t as $n \to \infty$.*

(iv) *If a sequence of continuous functions $\psi_n : [0,+\infty) \to \mathbf{R}$ satisfying $\lim_{x\to\infty} \psi_n(x) = 0$ uniformly converges to a continuous function $\psi : [0,+\infty) \to \mathbf{R}$ satisfying $\lim_{x\to\infty} \psi(x) = 0$, then $\psi_n(\mathcal{L}_n)$ strongly converges to $\psi(\mathcal{L})$ as $n \to \infty$.*

(v) *For all α, β ($\alpha < \beta$) which do not belong to the point spectrum of $\mathcal{L}$, $E_n((\alpha,\beta])$ strongly converges to $E((\alpha,\beta])$ as $n \to \infty$.*

(2) *Let $\sigma(\mathcal{L}_n)$ and $\sigma(\mathcal{L})$ be respectively the spectrum of $\mathcal{L}_n$ and $\mathcal{L}$. Then $\sigma(\mathcal{L}) \subset \lim_{n\to\infty} \sigma(\mathcal{L}_n)$; that is, for any $\lambda \in \sigma(\mathcal{L})$, there exists a sequence of $\lambda_n \in \sigma(\mathcal{L}_n)$ such that λ_n tends to λ as $n \to \infty$. Suppose moreover that $\{\mathcal{E}_n\}$ is asymptotically compact. Then the resolvent $G_{n;\zeta}$ and the semigroup $P_{n;t}$ are compact operators (consequently $\sigma(\mathcal{L}_n)$ is discrete), and for any $a, b \in \mathbf{R} \setminus \sigma(\mathcal{L})$ ($a < b$) we have $\dim E_n((a,b])H_n = \dim E((a,b])H$ for large n, and $\sigma(\mathcal{L}_n)$ converges to $\sigma(\mathcal{L})$ as $n \to \infty$.*

The definitions (4) and (5) above are in Mosco [**38**] for the case where all H_n coincide with H and Φ_n is the identity map of H, and the present ones are in Kuwae and Shioya [**34**].

Now in what follows, we would like to discuss the Mosco convergence of a sequence of Dirichlet spaces, and hence the convergence of the spectral structures as in Theorem 3.1. To begin with, we consider a regular Dirichlet space $X = (X, \mu, \mathcal{E})$ with the following properties:

[I] For some constants $\nu > 2$ and $A > 0$, the Sobolev inequality holds on X:

$$\left(\int_X |u|^{2\nu/(\nu-2)}\, d\mu\right)^{(\nu-2)/\nu} \leq A'\left\{\mathcal{E}(u) + \|u\|_{L^2}^2\right\}, \quad u \in D[\mathcal{E}_X].$$

[II] For a positive constant B, $\mu(X) \leq B$.

[III] $1 \in D[\mathcal{E}]$ and $\mathcal{E}(1,1) = 0$.

THEOREM 3.2. *Let $X_n = (X_n, \mu_n, \mathcal{E}_n)$ be a sequence of regular Dirichlet spaces satisfying the properties [I], [II], and [III] with some constants $\nu > 2$, $A > 0$, and $B > 0$. Then there exist a subsequence $\{X_m\}$, a regular Dirichlet space $(X, \mu, \mathcal{E})$ with the same properties, and a linear map $\Phi_m : \mathcal{C} \to L^2(X_m, \mu_m)$ from a core $\mathcal{C}$ of $\mathcal{E}$ into $L^2(X_m, \mu_m)$ satisfying [*] such that $\mathcal{E}_m$ converges to $\mathcal{E}$ in the sense of Mosco as $m \to \infty$. Moreover, $\{\mathcal{E}_m\}$ is asymptotically compact.*

The properties [I] and [II] enable us to embed regular Dirichlet spaces into a fixed compact metric space, and it follows from [III] that the algebra generated by the eigenfunctions of a given regular Dirichlet space contains the unit element 1. In fact, Theorem 3.2 can be proved as follows: Let $X = (X, \mu, \mathcal{E})$ be a regular Dirichlet space as above. Then in view of property [I], the semigroup possesses an integral kernel $p(t,x,y)$ satisfying $p(t,x,y) \leq A_* r^{-\nu/2}$ for almost all $x, y \in X$ and $0 < t \leq 1$, where A_* is a constant depending only on A and ν. In addition, property [II] allows us to have a complete orthonormal system of eigenfunctions ϕ_i of the eigenvalue λ_i and the expansion of the kernel, that is, $p(t,x,y) = \sum_{i=0}^{\infty} e^{-\lambda_i t}\phi_i(x)\phi_i(y)$. We remark that $\lambda_i \geq CB^{-1}(i+1)^{2/\nu}$ for $\lambda_i \geq 1$ and $i+1 \leq BC$ for $\lambda_i \leq 1$; in addition, $\sum_{i=0}^{\infty} \lambda_i^{\nu} e^{-t\lambda_i}\phi_i(x)^2 \leq C(t^{-3\nu/2} + t^{-\nu})$ for all $x \in X$ and $0 < t < +\infty$, where C is a positive constant depending only on ν and A. By setting

$$\gamma_x(t) = (e^{-(t+1/t)/2} e^{-\lambda_i t/2}\phi_i(x))_{i=0,1,2,\ldots}, \quad x \in X,\ 0 \leq t \leq +\infty,$$

we can assign points of X continuous loops in the Hilbert space $\ell^2 = \{(a_i)_{i=0,1,2,\ldots} \mid \sum_{i=0}^{\infty} {a_i}^2 < +\infty\}$ and assume that X is a subset of the Banach space consisting of continuous loops γ at the origin of ℓ^2 endowed with the norm

$$\|\gamma\| = \sup_{0 \leq t \leq +\infty} |\gamma(t)|_{\ell^2}.$$

Then from the estimates on λ_i and ϕ_i, we see that the subset is included in a compact subspace $K(\nu, A, B)$ depending only on the given constants. Let us introduce a distance on X by

$$d_X^{spec}(x,y) = \left(\sup_{t>0} e^{-(t+1/t)}(p(t,x,x) + p(t,y,y) - 2p(t,x,y))\right)^{1/2}, \quad x, y \in X.$$

Then we have $d_X^{spec}(x,y) = \|\gamma_x - \gamma_y\|$, $x, y \in X$, and hence (X, d_X^{spec}) can be realized as a subspace in $K(\nu, A, B)$. Therefore a sequence $\{X_n\}$ in Theorem 3.2 contains a subsequence $\{X_m\}$ which converges to a compact subspace $\widetilde{X}$ of $K(\nu, A, B)$ as $m \to \infty$ with respect to the Hausdorff distance in this metric space. Thus there exist a sequence of positive numbers ε_m tending to 0 as $m \to \infty$, Borel measurable maps $f_m : X_m \to \widetilde{X}$, and a nonnegative Radon measure μ on $\widetilde{X}$ such that

$|d^{spec}_{X_m}(x,y) - d_{\widetilde{X}}(f_m(x), f_m(y))| < \varepsilon_m$ for all $x, y \in X_m$, $\widetilde{X} \subset (f_m(X_m))_{\varepsilon_m}$, and $f_{m*}\mu_m$ vaguely converges to μ as $m \to \infty$. Moreover there exist a sequence of continuous functions ψ_i $(i = 0, 1, 2, \dots)$ on $\widetilde{X}$ and a sequence of numbers $\{\lambda_i \mid 0 = \lambda_0 \le \lambda_1 \le \lambda_2 \le \cdots (\le \infty)\}$ such that each element γ of $\widetilde{X}$ can be expressed as $\gamma = (e^{-(t+1/t)} e^{-\lambda_i t} \psi_i(\gamma))_{i=0,1,2,\dots}$. Then it follows from property [III] that the algebra $\mathcal{C}$ generated by the system $\{\psi_i\}$ separates points of $\widetilde{X}$ and contains the unit element 1. This implies that $\mathcal{C}$ is dense in the Banach space $C(\widetilde{X})$. Now if we denote the support of the limit measure μ by X, then we can deduce that $\{\psi_i\}$ is a complete orthonormal system of eigenfunctions of the eigenvalues λ_i with respect to a regular Dirichlet form $\mathcal{E}$ on $L^2(X, \mu)$. The linear map $\Phi_m : \mathcal{C} \to L^2(X_m, \mu_m)$ in Theorem 3.2 is now given by

$$\Phi_m(u) = u \circ f_m, \quad u \in \mathcal{C};$$

furthermore, the approximation $f_m : X_m \to \widetilde{X}$ satisfies

$$\sup_{t>0} e^{-(t+1/t)} |p_{X_m}(t, x, y) - p(t, f_m(x), f_m(y))| \le \varepsilon_m, \quad x, y \in X_m,$$

where $p(t, x, y)$ denotes the continuous kernel of the semigroup associated with the form $\mathcal{E}$. In this sense, the map f_m is spectral approximation between X_m and X.

We refer the reader to [**26**], Part I, [**28**] or [**30**] for details on Theorem 3.2. In [**30**], the arguments are based on a distance introduced on the set of regular Dirichlet spaces satisfying [I], [II], and [III]. See also [**3**], which motivates the introduction of the distance in [**30**].

Let us now consider strongly local, regular Dirichlet spaces which satisfy [I], [II], [III], and also [C] in Theorem 2.1. Then we have

$$d_{\mathcal{E}}(x, y) \le \theta(d^{spec}_X(x, y))$$

for all $x, y \in X$ and

$$|d_{\mathcal{E}}(x, y) - d_{\mathcal{E}}(x', y')| \le \theta(d^{spec}_X(x, y) + d^{spec}_X(x', y'))$$

for all $x, x', y, y' \in X$, where $\theta(t)$ is a monotone increasing continuous function with $\theta(0) = 0$ depending only on the given constants ν, A and B. This suggests that the approximating map f_m would also be such a map for the intrinsic distance $d_{\mathcal{E}_m}$ of X_m in Theorem 3.2. In fact, we have the following.

THEOREM 3.3. *Let $(X_m, \mu_m, \mathcal{E}_m)$ and $(X, \mu, \mathcal{E})$ be as in Theorem 3.2 and let $f_m : X_m \to \widetilde{X}$ be the approximating maps described as above. Suppose that X_m is strongly local and satisfies property [C]. Then the assertion below holds:*

(1) A subsequence of $\{f_m\}$, denoted again by the same letter, gives an ε_m-Hausdorff approximation between the intrinsic distance $d_{\mathcal{E}_m}$ on X_m and a continuous pseudo-distance δ on $\widetilde{X}$, where ε_m tends to zero as $m \to \infty$. In addition, $\delta(x, X) = 0$ for any $x \in \widetilde{X}$.

Suppose further that X_m satisfies the volume doubling property $[\mathrm{D}]_{\kappa_m}$ for some constants κ_m and R_m. Let δ be a continuous pseudo-distance on $\widetilde{X}$ obtained as above. Then the following hold:

(2) $C^{0,1}(X, \delta) \subset D[\mathcal{E}]$, and for any $u \in C^{0,1}(X, \delta)$ the energy measure $\mu_{\langle u,u \rangle}$ is absolutely continuous with respect to the measure μ and its density $\Gamma(u, u)$ satisfies

$$\gamma(u, u)^{1/2} \le \mathrm{dil}_\delta\, u \quad \mu\text{-a.e.}$$

(3) *If δ is nondegenerate in an open subset V of X, then the part of $\mathcal{E}$ on V is strongly local.*

(4) *Let $\{V_m\}$ be a sequence of open subsets of X_m which converges to an open subset V of X. If the volume doubling property $[\mathrm{D}]_\kappa$ and the weak Poincaré inequality $[\mathrm{P}]_{\mathcal{E}_m,\tau}$ are satisfied on V_m for some positive constants κ and τ, then the pseudo-distance δ is nondegenerate on V and the same properties $[\mathrm{D}]_\kappa$ and $[\mathrm{P}]_{\mathcal{E},\tau}$ with respect to δ hold on V; moreover, for any $u \in C^{0,1}_{loc}(V, \delta)$,*

$$C \operatorname{dil}_\delta u \leq \Gamma(u,u)^{1/2} \quad \mu\text{-a.e.},$$

where C is a positive constant depending only on κ and τ.

Let us denote by X_δ the quotient space obtained by identifying points x and y of $\widetilde{X}$ with $\delta(x,y) = 0$. Then (X_δ, δ) is the Gromov-Hausdorff limit of the metric spaces $(X_m, d_{\mathcal{E}_m})$. Even if all $\mathcal{E}_m$ are strongly local, this is not the case for the limit form $\mathcal{E}$ in general (cf. [**32**]). In the case where the limit form is nonlocal, δ must be degenerate somewhere. On the other hand, we have examples such that $\mathcal{E}$ is strongly local but δ is identically zero; namely, X_δ is a single point. The assertion (4) implies, however, that under the conditions $[\mathrm{D}]_\kappa$ and $[\mathrm{P}]_{\mathcal{E},\tau}$, the limit spaces $\widetilde{X}$, X, and X_δ coincide. We refer the reader to [**26**], Part I, or [**28**] for the proof of Theorem 3.3 and examples illustrating the roles of the conditions. See also, e.g., [**1**], [**31**], [**50**] for other related results.

Let us consider a family $\mathcal{F}$ of pointed, strongly local, regular Dirichlet spaces $X = (X, o, \mu, \mathcal{E})$ with the following properties:

[IV] Each space $X \in \mathcal{F}$ satisfies [C], and the intrinsic distance $d_\mathcal{E}$ is complete.

[V] There exist positive constants κ, τ and a positive nondecreasing function $\rho(r)$ $(0 \leq r < +\infty)$ such that for any $X \in \mathcal{F}$, the volume doubling property $[\mathrm{D}]_\kappa$ and the weak Poincaré inequality $[\mathrm{P}]_{\mathcal{E},\tau}$ hold in the metric balls $B(x, \rho(d_\mathcal{E}(o,x)))$ centered at $x \in X$ of radius $\rho(d_\mathcal{E}(o,x))$.

[VI] $\mu(B(o,1)) = 1$ for all $X \in \mathcal{F}$.

Given an element $X = (X, o, \mu, \mathcal{E}, d_\mathcal{E}))$ in $\mathcal{F}$, we can apply Theorem 2.3 to obtain the following:

$$\mu((B(o,R)) \leq \max\{1, (2\rho(0))^\kappa\} \exp C(1 + R/\rho(R));$$

$$\mu(B(x,\varepsilon)) \geq \left(\frac{2\varepsilon}{\rho(R)}\right)^\kappa \exp -C(1 + 2R/\rho(R)), \quad B(x,\varepsilon) \subset B(o,R),$$

where C is a positive constant depending only on κ and τ. Using these estimates on the volume growth of metric balls, we get an upper bound for the cardinality of ε-separated sets, and in view of Theorem 1.1 (1), we can conclude that any sequence in $\mathcal{F}$ contains a subsequence which converges in the Gromov-Hausdorff sense.

We make another observation on an element X in $\mathcal{F}$. Let us denote by $\mathcal{E}_R$ the part of $\mathcal{E}$ on the metric ball $B(o,R)$ centered at the reference point o of radius R. Let $p_R(t,x,y)$ be the (heat) kernel associated with $\mathcal{E}_R$, and take a complete orthonormal system $\{\psi_{R,i} \mid i = 1, 2, \ldots\}$ of eigenfunctions of $\mathcal{E}_R$. Then for any $R' < R$, we can get uniform Hölder estimates of $p_R(t,x,y)$ on $B(o,R') \times B(o,R')$ and also of $\psi_{R,i}$ on $B(o,R')$, where the estimates can be expressed in terms of the intrinsic distance $d_\mathcal{E}$. In fact, the conditions [IV], [V], and [VI] enable us to run the Moser iteration method.

These observations allow us to deduce the following.

THEOREM 3.4. *Given a sequence* $X_n = (X_n, o_n, \mu_n, \mathcal{E}_n, d_n)$ $(d_n = d_{\mathcal{E}_n})$ *in the family* $\mathcal{F}$ *as above, one has a subsequence, denoted by the same letters, and a pointed, strongly local, regular Dirichlet space* $X = (X, o, \mu, \mathcal{E}, \delta)$ *with a distance* δ *which induces the same topology of* X, *such that the following assertions hold:*

(1) (X_n, o_n, μ_n, d_n) *converges to* (X, o, μ, δ) *in the measured Gromov-Hausdorff sense via* ε_n*-Hausdorff approximation* $f_n : B(o_n, R_n) \to X$, *where* ε_n *tends to* 0 *and* R_n *diverges as* $n \to \infty$.

(2) *The sequence of the Dirichlet forms* $\mathcal{E}_n$ *converges to* $\mathcal{E}$ *in the sense of Mosco via the approximating maps* f_n.

(3) *The limit space* X *also satisfies properties [IV] and [VI], and moreover*

$$C\ d_{\mathcal{E}} \leq \delta \leq d_{\mathcal{E}},\ C\ \mathrm{dil}_\delta u \leq \Gamma(u,u)^{1/2} \leq \mathrm{dil}_\delta u, \quad u \in C^{0,1}_{loc}(X, \delta)$$

for some positive constant C *depending only on* κ *and* τ.

(4) *The limit space* X *has property [V] with respect to the distance* δ.

Now we close this section with the following statement of a result due to Cheeger and Colding [**9**].

THEOREM 3.5. *Let* $M_n = (M_n, o_n, d_n, \mu_n)$ *be a sequence of complete, pointed Riemannian manifolds of dimension* k *which converges to a pointed metric measure space* $X = (X, o, \delta, \mu)$ *in the measured Gromov-Hausdorff sense, where the Riemannian measure* μ_n *of* M_n *is normalized by* $\mu_n(B(o_n, 1)) = 1$. *Suppose there exists a nonnegative nondecreasing function* $\xi(r)$ *such that the Ricci curvature of* M_n *at any point* x *is bounded from below by* $-(k-1)\xi(d_n(o_n, x))$. *Then the energy form* $\mathcal{E}_n$ *of* M_n *converges to the Dirichlet* 2*-energy* $\mathcal{E}$ *of* X *in the sense of Mosco, that is, the energy form defined by*

$$\mathcal{E}(u,u) = \int_X (\mathrm{dil}_\delta u)^2\, d\mu, \quad u \in C^{0,1}_{loc}(X, \delta) \cap D[\mathcal{E}].$$

In fact, this is derived from deep analysis on the approximating manifolds and the limit space based on the volume doubling property, the Poincaré inequality, and the Bochner technique.

4. Convergence of Riemannian vector bundles and energy of maps

In this section, we discuss convergence on the energy forms of Riemannian vector bundles and also on the energy functionals on maps.

4.1. Let us consider a Riemannian vector bundle V on a compact Riemannian manifold (M, g), that is, a complex vector bundle equipped with an Hermitian metric $(,)$ and a compatible connection ∇. In addition, we are given a continuous section R of the endomorphism bundle of V. Then we have a (densely defined) closed form $\mathcal{F}_{V,R}$ on the Hilbert space $L^2(M, V)$ of L^2 sections of the vector bundle which is defined by

$$\mathcal{F}_{V,R}(\sigma) = \mathcal{F}_{V,R}(\sigma, \sigma) = \int_M |\nabla \sigma|^2 + (R(\sigma), \sigma)\ \mu_M, \quad \sigma \in D[\mathcal{F}_{V,R}].$$

In the case where $R = 0$, we write $\mathcal{F}_V$ for $\mathcal{F}_{V,0}$. In this case, we recall Kato's inequality

$$\mathcal{E}_M(|\sigma|) \leq \mathcal{F}_V(\sigma), \quad \sigma \in D[\mathcal{F}_V]$$

(see [**23**]).

In what follows, we are given, as above, a sequence of Riemannian vector bundles (M_n, V_n, R_n), and we state a result on its Mosco convergence. Here we assume that the base manifolds are equipped with the Riemannian measures μ_n normalized by $\mu_n(M_n) = 1$, and furthermore the ranks of the vector bundles are equal to a given constant k.

THEOREM 4.1. *Let (M_n, V_n, R_n) be as above. Suppose that all M_n satisfy the Sobolev inequality [S] with some constants $\nu > 2$ and $A > 0$, and suppose that $\|r_n^-\|_{L^p} \leq K$ for some constants $p > \nu/2$ and $K > 0$ or $\|r_n^-\|_{L^{\nu/2}} \leq \alpha/A$ for some constant $\alpha \in (0,1)$, where $r_n^-(x) = \max\{0, -r_n(x)\}$ $(x \in M_n)$ and $r_n(x)$ stands for the minimum of the eigenvalues of the symmetric linear map $R_x \in \mathrm{Hom}(V_{n|x}, V_{n|x})$. In the case where the least eigenvalue of $\mathcal{F}_{V_n,R_n}$ does not diverge to infinity as $n \to \infty$, there exists a subsequence $(M_m, V_m.R_m)$, a (densely defined) closed form $\mathcal{F}_\infty$ on the Hilbert space of L^2 sections of a continuous Hermitian vector bundle V_∞ of rank k over a locally compact, separable metric space Y endowed with a Radon measure μ_Y, and a linear map Φ_n from a dense subspace $\mathcal{C}$ of the Hilbert space $L^2(Y, V_\infty)$ into $L^2(M_n, V)$ with the property [*] such that $\{\mathcal{F}_{V_n,R_n}\}$ is asymptotically compact and it converges to $\mathcal{F}_\infty$ in the sense of Mosco as $m \to \infty$.*

Let us outline the proof of this theorem. See [**29**] for details. First, for each n, we consider the principal bundle U_n of unitary frames of V_n with the structure group $U(k)$, where $U(k)$ denotes the unitary group. Then sections of V_n correspond to functions from U_n to $\mathbf{C}^k$ which are equivariant under the action of $U(k)$. In this correspondence, the quadratic form $\mathcal{F}_{V_n}$ is replaced by the energy form on the subspace of equivariant functions with respect to the Riemannian metric $\widetilde{g}_n$ on U_n which is naturally induced from the Riemannian metric of M_n and the normalized bi-invarinat metric of $U(k)$. On the other hand, we recall that $(U_n, \widetilde{g}_n)$ also satisfies the Sobolev inequality [S] with constants $\nu' = \nu + k^* > 2$ $(k^* = \dim U(k))$ and A' which depend only on A and k. Hence we can find a subsequence $\{U_m\}$ such that U_m converges to a regular Dirichlet space $(\widetilde{Y}, \widetilde{\mu}, \widetilde{\mathcal{E}})$ as $m \to \infty$ in the sense of Mosco. Moreover, it turns out that the unitary group $U(k)$ acts on the limit space $\widetilde{Y}$ in such a way that the action is continuous, it keeps the measure $\widetilde{\mu}$ invariant, and the approximation maps from U_m to $\widetilde{Y}$ are asymptotically equivariant; in addition, the action of $U(k)$ on $\widetilde{Y}$ is free on an open subset $\hat{Y}$ of $\widetilde{Y}$. Thus the action yields a Radon measure μ_Y on the quotient space $Y = \hat{Y}/U(k)$ and a continuous Hermitian vector bundle V_∞ over Y, and also the limit Dirichlet form $\widetilde{\mathcal{E}}$ induces a closed form $\mathcal{F}_\infty^{(0)}$ on the Hilbert space of L^2 sections of V_∞. We note that the eigensections of the form $\mathcal{F}_\infty^{(0)}$ are all continuous, and moreover we remark that the sequence of the base manifolds M_m may be assumed to converge to a regular Dirichlet space $(X, \mu_X, \mathcal{E}_X)$ as $m \to \infty$ in the sense of Mosco. Then we have a surjective and continuous map π of $\widetilde{Y}$ onto X such that the push-forward of the measure $\widetilde{\mu}$ coincides with μ_X and the pull-backs of continuous functions on X belong to the subspace of $U(k)$-invariant continuous functions on $\widetilde{Y}$. However, the converse is not true in general; for instance, the continuous function $|\sigma|$ on $\hat{Y}$ induced by a continuous section σ of V_∞ over Y can be assumed to be defined on $\widetilde{Y}$ by setting $|\sigma| = 0$ outside of $\hat{Y}$, but the function does not come from continuous functions on X in general. Second, we consider the forms $\{\mathcal{F}_{V_m,R_m}\}$. Then in view of Kato's inequality and the conditions on R_m, we can deduce that for some constants $\alpha \in (0,1)$ and $\beta > 0$ and for all m,

$$\alpha \mathcal{F}_{V_n}(\sigma) - \beta \|\sigma\|_{L^2} \leq \mathcal{F}_{V_m,R_m}(\sigma),$$

$\sigma \in D[\mathcal{F}_{V_m}]$. This shows that the forms are asymptotically compact, and hence by taking a subsequence if necessary, we can conclude that $\{\mathcal{F}_{V_m,R_m}\}$ converges to a closed form $\mathcal{F}_\infty$ with $D[\mathcal{F}_\infty] \subset D[\mathcal{F}_\infty^{(0)}]$ in the sense of Mosco.

In geometry and analysis on manifolds, we have important operators associated with quadratic forms $\mathcal{F}_{V,R}$ such as the Hodge Laplacian $\bar{\Delta} = \nabla^*\nabla + \mathcal{R}$ acting on differential p-forms on a compact Riemannian manifold M which is associated with the energy form $\int_M |d\omega|^2 + |\delta\omega|^2 d\mu_M$; in case $p = 1$, $\mathcal{R}$ is the Ricci curvature of M. See [**36**] for related results.

4.2. In this part, we consider a strongly local, regular Dirichlet space $(X, \mu, \mathcal{E})$ satisfying property [C], and a complete Riemannian manifold $N = (N, g_N)$, and we study the Sobolev space of maps from X to N and the energy functional explained below. Let us first assume for convenience that N is isometrically embedded into some Euclidean space $\mathbf{R}^D$, and denote the embedding by $\iota = (\iota^1, \ldots, \iota^D) : N \to \mathbf{R}^D$. For a map $u : X \to N$, we write $\iota \circ u = (u^1, \ldots, u^D)$. The Sobolev space $W^{1,2}(X, N)$ consists of maps u such that each component u^i $(i = 1, \ldots, D)$ belongs to the domain $D[\mathcal{E}]$, and $u(x) = (u^1(x), \ldots, u^D(x)) \in N$ for almost all $x \in X$. Furthermore, given $u \in W^{1,2}(X, N)$, the energy measure $\mu^N_{\langle u \rangle}$ and the energy of u can be defined by $\mu^N_{\langle u \rangle} = \sum_{i=1}^D \mu_{\langle u^i, u^i \rangle}$ and $\mathcal{E}_{X,N}(u) = \int_X \mu^N_{\langle u \rangle}$, respectively. We notice that if $u \in W^{1,2}(X, N)$ is continuous, then

$$\begin{aligned}
\mathcal{E}_{X,N}(u) &= \lim_{t \to 0} \int\int_{X \times X} \sum_{i=1}^D (u^i(x) - u^i(y))^2 \frac{1}{t} p(t, x, y)\, d\mu(x) d\mu(y) \\
&= \lim_{t \to 0} \int\int_{X \times X} \frac{d_N(u(x), u(y))^2}{d_{\mathcal{E}}(x, y)^2} \frac{d_{\mathcal{E}}(x, y)^2}{t} p(t, x, y)\, d\mu(x) d\mu(y),
\end{aligned}$$

where $p(t, x, y)$ and d_N respectively stand for the (heat) kernel of $\mathcal{E}$ and the Riemannian distance of N.

In what follows, we assume, unless otherwise stated, that X and N are both compact, and we consider homotopy classes of continuous maps $C(X, N)$ from X to N. Then $C(X, N) \cap W^{1,2}(X, N)$ is dense in $C(X, N)$ and any $u \in C(X, N)$ is homotopic to some $v \in C(X, N) \cap W^{1,2}(X, N)$. For this reason, we set $\sigma_{X,N}(A) = \inf\{\mathcal{E}_{X,N}(u) \mid u \in A \cap W^{1,2}(X, N)\}$ for a homotopy class A, and we denote by $\Sigma_{X,N}(A)$ the set of $u \in A \cap W^{1,2}(X, N)$ satisfying $\mathcal{E}_{X,N}(u) = \sigma_{X,N}(A)$. We recall here that even if the domain is a Riemannian manifold M, $\Sigma_{M,N}(A)$ could be empty; however, if the target manifold N has nonpositive curvature, then the classical result by J. Eells and J. H. Sampson asserts that $\Sigma_{M,N}(A) \neq \emptyset$. Moreover, recent studies by Eells and Fuglede [**12**] show that a compact Riemannian polyhedron P which is admissible in their sense satisfies $\Sigma_{P,N}(A) \neq \emptyset$ (see also [**13**], [**14**]). In fact, for such a space P, there exist positive constants R, κ, and τ such that in $B(x, R)$, the volume doubling property $[\mathrm{D}]_\kappa$ and the weak Poincaré inequality $[P]_{\mathcal{E},\tau}$ hold, and further, any $u \in \Sigma_{P,N}(A)$ is a Hölder continuous map with exponent $\alpha \in (0,1)$ satisfying

$$[\mathrm{H}] \qquad d_N(u(x), u(y)) \leq CR \left(\fint_{B(z,R)} \mu_{\langle u \rangle} \right)^{1/2} \left(\frac{d_{\mathcal{E}}(x, y)}{R} \right)^\alpha$$

for all $x, y \in B(z, R/10) \subset P$, where C and α are positive constants depending only on the given κ and τ.

In the sequel, we assume that the Dirichlet space $X = (X, \mu, \mathcal{E})$ also possesses properties $[D]_\kappa$ and $[P]_{\mathcal{E},\tau}$, and the target manifold N is nonpositively curved. In this case, for $u, v \in C(X, N) \cap W^{1,2}(X, N)$, if u and v are homotopic, then they are homotopic in $C(X, N) \cap W^{1,2}(X, N)$, and for a homotopy class A, if $\Sigma_{X,N}(A) \neq \emptyset$, then the geodesic homotopy $\{u_t \mid 0 \leq t \leq 1\} \subset A$ joining two maps $u_0, u_1 \in A$ stays in $\Sigma_{X,N}(A)$, and $\mu^N_{\langle u_t \rangle} = t\mu^N_{\langle u_1 \rangle} + (1-t)\mu^N_{\langle u_0 \rangle}$. These facts are consequences of the definition of the energy of continuous maps as above and the convexity of the universal covering space of N. Note however that the Euler-Lagrange equation for $u \in \Sigma_{X,N}(A)$ must be expressed in terms of the signed measures.

We would like to ask whether the existence of the universal covering space of the domain X would be sufficient to assert that $\Sigma_{X,N}(A) \neq \emptyset$ for all homotopy classes A.

Let us now restrict our attention to a family $\mathcal{F}'$ of compact, admissible Riemannian polyhedra uniformly satisfying $[\mathrm{D}]_\kappa$ and $[\mathrm{P}]_{\mathcal{E},\tau}$. Moreover, we assume that $\mu(B(x, R)) \geq \theta$ for some positive constant θ and for all $x \in X, X \in \mathcal{F}$. Then for $X \in \mathcal{F}'$, it can be deduced from [H] that the cardinality of the homotopy classes $A \in [X, N]$ with $\sigma(A) \leq \lambda$ is finite for each $\lambda > 0$, so that the set of homotopy classes $[X, N]$ can be put in order as $0 = \sigma_{X,N}(A_0) < \sigma_{X,N}(A_1) \leq \sigma_{X,N}(A_3) \leq \cdots$. We set $\sigma_i(X, N) = \sigma_{X,N}(A_i)$, and if $[X, N]$ consists of finite classes and we denote by $\nu(X, N)$ its cardinality, then we understand that $\sigma_i(X, N) = +\infty$ for $i > \nu(X, N)$ for convenience.

Now we are in a position to state the following.

THEOREM 4.2. *Let $X_n = (X_n, \mu_n, \mathcal{E}_n)$ be a sequence of a family $\mathcal{F}'$ of compact, admissible Riemannian polyhedra satisfying the properties described above with given positive constants R, κ, τ, and θ. Suppose that $\{X_n\}$ converges to a regular Dirichlet space $X = (X, \mu, \mathcal{E})$ (via approximating maps $f_n : X_n \to X$) in the sense of Mosco. Then X also satisfies $\Sigma_{X,N}(A) \neq \emptyset$ for all homotopy classes $A \in [X, N]$; moreover, for each $i = 0, 1, 2, \ldots$,*

$$\lim_{n\to\infty} \sigma_i(X_n, N) = \sigma_i(X, N),$$

and for any $\phi \in \Sigma_{X,N}(A_{n,i})$, there exists a map $\psi \in \Sigma_{X,N}(A_i)$ such that

$$|\phi - \psi \circ f_n|_{L^\infty} \leq \varepsilon_{i;n},$$

where $\{\varepsilon_{i;n}\}$ is a sequence of positive numbers tending to 0 as $n \to \infty$.

Let us now consider the case where $\dim N = 1$, that is, $N = \mathbf{R}/\mathbf{Z}$. Then the set of homotopy classes $[X, \mathbf{R}/\mathbf{Z}]$ naturally becomes an abelian group. In addition, if the domain is a compact Riemann manifold M, then $[M, \mathbf{R}/\mathbf{Z}]$ corresponds to the abelian group of harmonic 1-forms of integral periods on M. This yields a harmonic map $\alpha_M : M \to \mathcal{A}_M$, called the Albanese map, from M to a flat torus $\mathcal{A}_M$ of dimension $b_1(M)$, called the Albanese torus of M, where $b_1(M)$ denotes the first Betti number of M.

THEOREM 4.3. *Let $\{M_n\}$ be a sequence of compact Riemannian manifolds satisfying the same properties described above with given positive constants R, κ, τ, and θ. Suppose that it converges to a regular Dirichlet space X in the sense of Mosco as $n \to \infty$ and that $b_1(M_n)$ is uniformly bounded from above. Then the* *sequence of the Albanese tori $\mathcal{A}_{M_n}$ converges to a flat torus T of dimension k, $0 \le k \le \liminf_{n\to\infty} b_1(M_n)$, both in the topology of the Gromov-Hausdorff distance and also in the sense of Mosco. Moreover, the Albanese map $\alpha_{M_n} : M_n \to \mathcal{A}_{M_n}$ uniformly converges to a Hölder continuous map $\alpha_\infty : X \to T$, and α_∞ has the least energy in the homotopy class.*

A family of compact Riemannian manifolds of the same dimension such that the diameters are uniformly bounded from above and the Ricci curvatures are uniformly bounded from below satisfies all the above properties. We note that for limit spaces X of this family, maps in $\Sigma_{X,N}(A)$ are all Lipschitz continuous. In fact, we can apply the Bochner technique.

Let $(X, \mu, \mathcal{E})$ be a strongly local, regular Dirichlet space satisfying the properties [C], $[\mathrm{D}]_\kappa$, and $[\mathrm{P}]_{\mathcal{E},\tau}$, and Ω a relatively compact open set of X. We are concerned with the Dirichlet problem, the existence and regularity of maps of least energy with given boundary values. Let u be a map in $W^{1,2}(\Omega, N)$. We assume that the sectional curvature N is bounded from above by a nonnegative constant $\bar{K}^2$ and the injectivity radius of N at a point p is greater than a positive constant i. Then if $u(\Omega) \subset B(p,r)$ for some r, $0 < r < \min\{i, \pi/2\bar{K}\}$), we have a unique solution ϕ of the Dirichlet problem such that $\phi(\Omega) \subset B(p,r)$ and it is Hölder continuous in Ω. In fact, in this case, we can apply the direct method of the calculus of variations and the elliptic regularity arguments to weak solutions, because of the given map as boundary values being bounded in a convex region. However, as in Theorem 4.2, so far as the convergence of such solutions and their energies are concerned, we need certain restrictions on the convergence of domains and boundary values. We refer the reader to [**25**] and [**26**], Part II, for the details on Theorems 4.2 and 4.3. See also [**27**], [**33**], and [**40**] for related results.

To end this exposition, we briefly mention some expansions of domains and target spaces of maps. In [**12**], [**13**], and [**14**], the target spaces are taken to be complete Riemannian polyhedra of upper bounded curvature. Another approach to the extension of the domains and targets of maps and to the introduction of the notion of Mosco convergence on the spaces of maps has been tried recently in [**44**]. We also refer the reader to [**43**] and the references therein for recent progress on geometric analysis on Alexandrov spaces.

References

[1] K. Akutagawa, Convergence for Yamabe metrics of positive scalar curvature with integral bound on curvature, Pacific J. Math. **175** (1996), 239–258.

[2] A. Bellaïche, The tangent space in sub-Riemannian geometry, in *Sub-Riemannina Geometry* (ed. A. Bellaïche and J.-J. Risler), Progress in Mathematics **144**, Birkhäuser, Boston, 1–78.

[3] P. Bérard, G. Besson and S. Gallot, On embedding Riemannian manifolds in a Hilbert space using their heat kernels, Prépublication de l'Institut Fourier, 109, 1988; Embedding Riemannian manifolds by their heat kernels, Geom. Funct. Anal. **4** (1994), 373–398.

[4] M. Biroli and U. Mosco, A Saint-Venaut principle for Dirichlet forms on discontinuous media, Ann. Mat. Pura Appl. (4) **169** (1995), 125–181.

[5] M. Bourdon and H. Pajot, Poincaré inequalities and quasiconformal structure on the boundary of some hyperbolic buildings, Proc. Amer. Math. Soc., **127** (1999), 2315–2324.

[6] D. Burago, Y. Burago and S. Ivanov, A Course in Metric Geometry, GSM **33**, Amer. Math. Soc., Providence, R.I., 2001.

[7] P. Buser, A note on the isoperimetric constant, Ann. Sci. École Norm. Sup. **15** (1982), 213–230.

[8] J. Cheeger, Differentiability of Lipschitz functions on metric spaces, Geom. Funct. Anal. **9** (1999), 428–517.

[9] J. Cheeger and T.H. Colding, On the structure of spaces with Ricci curvature bounded below, I, J. Differential Geom. **46** (1997), 406–480 ; — II, *ibid.* **54** (2000), 13–35 : — III, *ibid.*, 37–74.
[10] E.B. Davies, Heat Kernels and Spectral Theory, Cambridge University Press, Cambridge, 1989.
[11] G. Dal Maso, An Introduction to Γ–Convergence, Progress in Nonlinear Differential Equations and Their Applications **8**, Birkhäuser, Boston, 1993.
[12] J. Eells and B. Fuglede, Harmonic Maps between Riemannian Polyhedra, Cambridge University Press, Cambridge, 2001.
[13] B. Fuglede, Hölder continuity of harmonic maps from Riemannian polyhedra to spaces of upper bounded curvature, Calc. Var. P. D. E. **16** (2003), 375–403.
[14] B. Fuglede, The Dirichlet problem for harmonic maps from Riemannian polyhedra togeodesic spaces of upper bounded curvature, Preprint Series 2001, No. 18, Dec. 2001, University of Copenhagen.
[15] K. Fukaya, Collapsing Riemannian manifolds and eigenvalues of the Laplace operator, Invent. Math. **87** (1987), 517–547
[16] K. Fukaya, Hausdorff convergence of Riemannian manifolds and its applications, in *Recent Topics in Differential and Analytic Geometry*, Adv. Stud. in Pure Math. **18** (1990), North-Holland, Amsterdam, and Kinokuniya, Tokyo, 143–238.
[17] M. Fukushima, Y. Oshima and M. Takeda, Dirichlet Forms and Symmetric Markov Processes, Walter de Gruyter, Berlin-New York, 1994.
[18] M. Gromov, Structures métriques pour les variétés riemanniennes, rédigé par J. LaFontaine et P. Pansu, Cedic Fernand-Nathan, Paris 1981.
[19] M. Gromov, Metric Structures for Riemannian and non-Riemannian spaces (with appendices by M. Katz, P. Pansu, and S. Semmes, ed. by J. LaFontaine and P. Pansu, English translation by S. M. Bates), Progress in Mathematics **152**, Birkhäuser, Boston, 1999.
[20] P. Hajłasz and P. Koskela, Sobolev met Poincaré, Mem. Amer. Math. Soc. **145** (2000), No. 688.
[21] B. Hanson and J. Heinonen, An n-dimensional space that admits a Poincaré inequality but has no manifold points, Proc. Amer. Math. Soc. **128** (2000), 3379–3390.
[22] J. Heinonen, Lectures on Analysis on Metric Spaces, Universitext, Springer-Verlag, New York, 2001
[23] H. Hess, R. Schrader and D. A. Uhlenbrock, Kato's inequality and the spectral distributions of Laplacians on compact Riemannian manifolds, J. Differential Geom. **15** (1980), 27–38.
[24] K. Ishige and M. Murata, Uniqueness of nonnegative solutions of the Cauchy problem for parabolic equations on manifolds or domains, Ann. Scuola Norm. Sup. Pisa **30** (2001), 171–223.
[25] A. Kasue, Convergence of Riemannian manifolds and harmonic maps, Math. preprint series, Osaka City University, 2000.
[26] A. Kasue, Convergence of Riemannian manifolds and Laplace operators, I, Ann. Institut Fourier **52** (2002), 1219–1257; —— II, to appear.
[27] A. Kasue, Convergence of Riemannian manifolds, Laplace operators and energy forms, in *Proceedings of the Fifth Pacific Rim Geometry Conference* (ed. S. Nishikawa), Tohoku Math. Publ. **20** (2001), 75–97.
[28] A. Kasue, Geometric analysis on metric measure spaces –differentiation of Lipschitz functions– (Japanese), to appear in Riemannian manifolds and their limits, Sūgaku Memoire, Math. Soc. of Japan.
[29] A. Kasue, Spectral convergence of Riemannian vector bundles (Japanese), Lecture Notes Series in Mathematics **17** (2002), Osaka University, 69–92.
[30] A. Kasue and H. Kumura, Spectral convergence of Riemannian manifolds, Tōhoku Math. J. **46** (1994), 147–179 ; ——— II, Tōhoku Math. J. **48** (1996), 71–120.
[31] A. Kasue and H. Kumura, Spectral convergence of conformally immersed surfaces with bounded mean curvature, J. Geom. Anal. **12** (2002).
[32] A. Kasue, H. Kumura and Y. Ogura, Convergence of heat kernels on a compact manifold, Kyushu J. Math. **51** (1997), 453–524.
[33] M. Kotani and T. Sunada, Albanese maps and off-diagonal long time asymptotics for the heat kernel, Comm. Math. Phys. **209** (2000), 633–670.
[34] K. Kuwae and T. Shioya, A convergence of spectral structures: a functional analytic theory and its applications to spectral geometry, to appear in Comm. Anal. Geom.

[35] T. Laakso, Ahlfors Q-regular spaces with arbitrary Q admitting weak Poincaré inequality, Geom. Funct. Anal. **10** (2000), 111–123.
[36] J. Lott, Collapsing and the differential form Laplacian: the case of a smooth limit space, Duke Math. J. **114** (2001), 267–306.
[37] X. Menguy, Noncollapsing examples with positive Ricci curvature and infinite topological type, Geom. Funct. Anal. **10** (2000), 600–627.
[38] U. Mosco, Composite media and asymptotic Dirichlet forms, J. Funct. Anal. **123** (1994), 368–421.
[39] H. Nakajima, A convergence theorem for Einstein metrics and ALE spaces (Japanese), Sūgaku **44** (1992), 133–146; English transl., Amer. Math. Soc. Transl. (2) **160** (1994), 79–94.
[40] Y. Ogura, Weak convergence of laws of stochastic processes on Riemannian manifolds, Probab. Theory Relat. Fields **119** (2001), 529–557.
[41] L. Saloff-Coste, A note on Poincaré, Sobolev and Harnack inequality, Duke Math. J. Internat. Math. Res. Notices **2** (1992), 27–38.
[42] T. Sakai, Riemannian manifolds with bounded Ricci curvature and their limits (Japanese), to appear in Riemannian manifolds and their limits, Sūgaku Memoire, Math. Soc. of Japan.
[43] T. Shioya, Analysis on Alexandrov spaces (Japanese), to appear in Riemannian manifolds and their limits, Sūgaku Memoire, Math. Soc. of Japan.
[44] T. Shioya, Convergence of metric spaces and (nonlinear) variational functionals (Japanese), in the abstracts of Conference on Differential Geometry, Tsukuba University, 2002.
[45] K. T. Sturm, On the geometry defined by Dirichlet forms, Seminar on Stochastic Processes, Random Fields and Applications (Ascona, 1993), E. Bolthausen et al., eds., Birkhäuser, Progress in Probability **36**, 1995, 231–242.
[46] K. T. Sturm, Analysis on local Dirichlet spaces III. The parabolic Harnack inequality, J. Math. Pures Appl. **75** (1996), 273–297
[47] K. T. Sturm, Is a Diffusion Process Determined by Its Intrinsic Metric?, Chaos Solitons Fractals **8** (1997), 1855–1890.
[48] T. Yamaguchi, The development of convergence theory for Riemannian manifolds (Japanese), Sūgaku **47** (1995), 46–61
[49] T. Yamaguchi, Collapsing Riemannian 4-manifolds (Japanese), Sūgaku **52** (2000), 172–186.
[50] K. Yoshikawa, Degeneration of algebraic manifolds and the continuity of the spectrum of the Laplacian, Nagoya Math. J. **146** (1997), 83–129.

Department of Mathematics, Kanazawa University, Kanazawa 920-1192, Japan
E-mail address: `kasue@kenroku.kanazawa-u.ac.jp`

Translated by ATSUSHI KASUE

Amer. Math. Soc. Transl.
(2) Vol. **215**, 2005

Subfactor theory and its applications: Operator algebras and quantum field theory

Yasuyuki Kawahigashi

ABSTRACT. We review subfactor theory and its applications to algebraic quantum field theory with emphasis on classification theory.

1. Brief introduction

My speciality is subfactor theory within the theory of operator algebras, and more specifically, I study mathematical structures appearing in operator algebraic approaches to algebraic quantum field theory. Everyone knows that the field of operator algebras exists, but my experience suggests that rather few know about concrete results and methods in the theory of operator algebras, beyond the appearance of its name in connection with noncommutative geometry and quantum invariants in low dimensional topology. Thus, taking this opportunity, I would like to start with explanations on our aims and methods in operator algebra theory for non-experts. After that, I will review my recent research results. I refrain from stating precise definitions and exact statements of theorems here, and refer the reader to the references for all such things.

2. Aims of the theory of operator algebras

The noncommutative geometry of Connes and quantum invariants in 3-dimensional topology, starting with the discovery of the Jones polynomial, are particularly famous among interactions of operator algebra theory and other fields of mathematics. My research is closely related to the latter, so I have written and talked about them on many occasions, but these are rather applications of operator algebra theory and are not exactly within operator algebra theory in the narrow sense. So before going into such applications, I would like to start with internal aims and techniques within operator algebra theory. I refer the reader to [**24**] as general textbooks on operator algebra theory.

We study algebras of bounded linear operators on a Hilbert space, which is usually assumed to be separable and infinite dimensional, in the theory of operator

This article originally appeared in Japanese in Sūgaku **54** (4) (2002), 337–347.
2000 *Mathematics Subject Classification.* Primary 46L37, 81T05; Secondary 46L60, 81T40.
Key words and phrases. Algebraic quantum field theory, subfactor, tensor category, Virasoro algebra.

algebras. We further require that an algebra be closed under the $*$-operation and an appropriate topology. Depending on the choice between the norm topology and the strong operator topology, we obtain a C^*-algebra and a von Neumann algebra. (The weak operator topology also gives a von Neumann algebra, and so does the strong operator topology.) The strong and weak operator topologies are weaker than the norm topology, and closedness under a weaker topology is a stronger condition, needless to say, so a von Neumann algebra is automatically a C^*-algebra, but the methods used to study C^*-algebras and von Neumann algebras are often technically different, so many operator algebraists think that the theory of operator algebras consists of two different, but not entirely disjoint, theories, C^*-algebras and von Neumann algebras. Of course there are several similarities and interactions between these two, so it is important to know what is happening in the other, and some people work really in both, but many people work in only one of them, and I work on von Neumann algebras. (Many people outside the theory of operator algebras think that operator algebra theory and C^*-algebra theory are synonyms, and this is not logically wrong in the above sense that a von Neumann algebra is automatically a C^*-algebra, but it runs against the feelings of many operator algebraists.) Here, note that it is very important to assume that an algebra is closed under an appropriate topology. For example, consider a discrete countable group G and its left regular representation λ. Linear combinations of the unitary operators λ_g, $g \in G$, make an infinite dimensional algebra, but we need to make a closure in an appropriate topology to obtain a C^*-algebra or a von Neumann algebra, in order to use a general theory of operator algebras. Some infinite dimensional algebras without completeness, or even a topology, have often been studied in other fields of mathematics, but I feel that, in general, a truly interesting structure emerges only after a completion. (It is true in many cases that we see interesting structures ready before a completion, and then we do not need to make a completion, but I feel that this is just because we happen to be in a lucky situation.)

The main aim in the theory of operator algebras is to understand the structures of C^*-algebras and von Neumann algebras, and thus classification theory is naturally a central topic. That is, we would like to obtain a (computable) complete invariant or a complete enumerative list for operator algebras themselves, their automorphisms, group actions on them, or their subalgebras. In this sense, it would be an ultimate goal to make a complete classification list of all the operator algebras; but such a classification seems to be hopelessly difficult, unfortunately, like the classification problems for all topological spaces or all Banach spaces. Needless to say, we have many other important problems besides the classification problems, such as investigating a property that holds for all or most operator algebras, determining the class of operator algebras for which a certain condition holds, computing (not necessarily complete, but interesting) invariants, and constructing intriguing examples. Voiculescu's free probability theory and the work on exact C^*-algebras by Kirchberg are particularly important in connection with such problems, but I limit this review to problems directly related to classifications, partly due to my limited taste and knowledge.

An important idea in classification theories of operator algebras is that those in a nice class can be completely classified by an algebraic invariant. Another equally important idea is that we can obtain a simple characterization on which algebraic invariants can indeed arise. These ideas originated in a series of stunning results by Alain Connes in the 1970s and have been sources of many studies. More detailed descriptions of these are given as follows. The conditions for "nice" classes are generally called "amenability". This notion originally appears for a discrete group G, and is defined as existence of a nontrivial positive and translation invariant linear functional on $\ell^\infty(G)$. (Here positivity of a linear functional means that it maps any positive elements in $\ell^\infty(G)$ to a positive number.) Abelian groups and finite groups are amenable. (However, even for the integer group $\mathbf{Z}$, such a linear functional is constructed only with a transcendental method arising from the axiom of choice. It often happens in operator algebra theory that such a transcendental map carries important pieces of information.) For general discrete groups, amenability is a condition "close" to commutativity in a sense. For example, free groups having at least two generators are "the most noncommutative" in the sense that elements never commute except for trivial cases, and they are typical examples of non-amenable groups. A notion of amenability is extended to non-discrete groups easily. We have a notion of amenability for operator algebras as an analogue of this amenability for groups, and this condition for operator algebras has many equivalent formulations. Connes' greatest achievement within operator algebra theory is a characterization of amenability for von Neumann algebras. That is, he has proved that amenability of a von Neumann algebra is equivalent to hyperfiniteness, which means that the algebra can be approximated internally by finite dimensional algebras. (Note that it is often a useful and important idea to pursue an analogue for operator algebras of a notion for groups or fields, based on the idea that groups, rings, and fields are all "similar".) Thus we have important problems, such as classification of amenable operator algebras, classification of actions of (not necessarily discrete) amenable groups on amenable operator algebras, and classification of amenable operator subalgebras (after a suitable definition of amenability for inclusions of operator algebras). In the last nearly thirty years, we have seen much great progress on such problems, but still we are far from the completion. The notion of amenability for groups may not be ubiquitous in all mathematics, but it is a decisively important condition in operator algebra theory. We have several reasons to believe that unified classification with a simple invariant would immediately fail, if we proceed beyond the amenable class. Also, from an operator algebraic viewpoint, an amenable discrete group is "almost like $\mathbf{Z}$", but the other discrete groups are "entirely different". (Discrete groups of a Lie group, such as $SL(n,\mathbf{Z})$, are now important objects in the theory of operator algebras, but they are very far from being amenable, and few operator algebraic results are known for them. For completely general operator algebras, it is believed that whatever strange condition we may think, *anything*, can happen, as long as it is not obviously prohibited. There is a so-called "Gromov principle" that any non-trivial statement for general discrete countable groups has a counterexample. Many statements on discrete groups can be translated into those operator algebras, by passing to group algebras, so a statement analogous to this principle is believed to hold for operator algebras.)

Next I explain what an "algebraic invariant" is, in general. By the word "algebraic", I mean that the same or a similar invariant can be defined without topology. We usually consider some maps, such as linear functionals on operator algebras, $*$-homomorphisms from one operator algebra to another, or more complicated maps, and study the set of all such maps with some equivalence relations. It is often the case that the set of equivalence classes is "small" and has some algebraic structure. A typical such example is K-theory for C^*-algebras. "Some equivalence relations" often arise from "trivialization" of inner automorphisms of an algebra, which is of the form $x \mapsto \mathrm{Ad}(u)(x) = uxu^*$ for some unitary u in the operator algebra. Examples of "some algebraic structure" are (finitely generated) abelian groups, tensor categories, and ergodic flows. (It may not be appropriate to call the last one "algebraic".) All the classification theorems on von Neumann algebras (and their subalgebras, group actions on them), initiated by Connes and developed by Haagerup, Jones, Ocneanu, Popa, and other people, claim that such algebraic invariants are complete under assumptions of some amenability. In the classification theory of simple separable C^*-algebras, the relevant algebraic invariants are K-theoretic, and many classification theories have been successful along the line of the Elliott conjecture stating that amenable simple separable C^*-algebras are completely classified by the K-theoretic invariants. (We are, however, still far from a complete solution to this conjecture.) Here, the K-theoretic invariants are essentially K_0- and K_1-groups, but we need some extra related data for the classification.

I also explain basic ideas that are commonly shared in this type of classification theories. A commutative von Neumann algebra is of a form $L^\infty(X, \mu)$ and a commutative (unital) C^*-algebra is of a form $C(X)$, where (X, μ) in the former is a measure space and X in the latter is a compact Hausdorff space. (If a commutative C^*-algebra is not unital, we need to consider continuous functions on a locally compact Hausdorff space vanishing at infinity.) Thus, it is an old idea that theories of von Neumann algebras and C^*-algebras are "noncommutative integration/ergodic theory" and "noncommutative topology", respectively. The noncommutative geometry of Connes is also a recent far-reaching extension of this idea. One important and rather surprising idea in classification theories mentioned above is that passing to noncommutative algebras makes classification statements simpler (although it does not make technical aspects easier at all). For example, consider classification of automorphisms of the hyperfinite II_1 factor, which has been one of the most fundamental examples of von Neumann algebras since the initiation of the theory by Murray and von Neumann, and is, of course, amenable. If we have two automorphisms α, β whose non-trivial powers are never inner, then we have another automorphism θ and a unitary u in the algebra so that we have $\mathrm{Ad}(u)\cdot\alpha = \theta\cdot\beta\cdot\theta^{-1}$. The inner automorphism $\mathrm{Ad}(u)$ is regarded as a "trivial automorphism", so that a "generic" automorphism, in the sense that no non-trivial powers become trivial, is essentially unique for this operator algebra. This is a great achievement of Connes. A commutative counterpart of an automorphism of the hyperfinite II_1 factor in ergodic theory is a measure-preserving ergodic transformation of a Lebesgue space with a probability measure, but a statement that all such transformations are conjugate is obviously false. (Actually such transformations are all orbit equivalent, and this fact is important for operator algebra theory, but orbit equivalence loses information about which group acts on a Lebesgue space, and this is a problem from the viewpoint of classification of group actions.) It is important that we have $\mathrm{Ad}(u)$ in the above classification theorem of Connes, and the fact that a noncommutative algebra can have an inner automorphism not equal to the identity map plays a role in uniqueness here. Furthermore, since commutative C^*-algebras should be amenable whichever definition we may use, a counterpart of the Elliott conjecture on K-theoretic classification of simple amenable C^*-algebras in the commutative setting would be a statement that (locally) compact Hausdorff spaces are classified up to homeomorphism with K-theory, which is, of course, entirely false. Recall that commutative C^*-algebras have "many" ideals (unless it is the scalar field $\mathbf{C}$), so they are far from being simple. Thus the Elliott conjecture predicts that we have a simple and plain classification statement under an assumption that C^*-algebras are close to being commutative in terms of amenability, but far from being commutative in terms of simplicity. It has been recently shown that even for amenable simple C^*-algebras, we have various rather strange phenomena, but still, in comparison to topology, we have a simpler picture after passing to the noncommutative world.

Finally, note that a proof of such a classification theorem is always highly technical and relies on improving approximations step by step while controlling errors.

3. Subfactor theory of Jones

Next, I explain how the subfactor theory initiated by V. F. R. Jones [**13**] fits into a general framework as above and why it is related to several "quantum somethings". I have written and talked on these topics on many occasions, so I will be rather brief and refer the reader to our book [**9**].

The definition of a simple algebra, of course, requires triviality of two-sided ideals (closed under an appropriate topology). For von Neumann algebras, it has been known that this condition is equivalent to the property that the center is equal to $\mathbf{C}$, and for a historic reason, the terminology "factor" has been used, rather than "simple von Neumann algebra". Since the days of Murray and von Neumann, studying factors has been a central theme in von Neumann algebras. Then in subfactor theory, we study the situation where a factor N is included in another factor M. Usually, the larger factor M is fixed, so we often write "a subfactor $N \subset M$", but we also use the terminology "a subfactor" to mean the inclusion of N in M in many cases. Again by the idea that groups, rings, and fields are all similar, we consider an analogue of an index of a subgroup and a degree of an extension of a field, and call it the Jones index of a subfactor. This Jones index takes a value in the interval $[1, \infty]$ and can be a non-integer. In this setting, we are interested in an analogue of the Galois theory of extensions of fields and its "quantization". (Usually, one considers a field and its extension in the theory of fields, and we can similarly consider here a factor and its extended factor, but in many cases, it is more natural to fix a larger factor and study its subfactors.) A direct analogue of classical Galois theory has been studied in the theory of operator algebras for many years, as follows. Consider a factor M and an action of a finite group G on M such that any non-trivial element of G acts as a non-inner automorphism of M. (Here the idea that an inner automorphism is "trivial" again appears.) We write M^G for the fixed point algebra of this action, and then it is automatically a factor and we can recover the group G just from the inclusion $M^G \subset G$. More concretely, we consider automorphisms of M that fix all the elements of the subfactor M^G, and then the original action of G appears. We also have a Galois correspondence between subgroups of G and intermediate algebras of M and M^G. The Jones index of this subfactor $M^G \subset M$ is the order of G. It is an important idea in subfactor theory to extend this viewpoint to a general subfactor $N \subset M$ and regard N as a "fixed point algebra" under an action of something like a group. Such an idea was materialized in a very strong form of paragroup theory by Ocneanu [**20**]. Here, I explain this theory in a form due to Longo and Izumi. Recall that if a subfactor indeed comes from an action of a finite group as a fixed point algebra, then the finite group can be recovered from the subfactor as a certain automorphism group of the larger algebra. For a general subfactor, we consider *endomorphisms* rather than automorphisms and obtain a certain algebraic system. (By simplicity of M, any endomorphism is automatically injective. It is more interesting to consider a non-surjective endomorphism. For such an endomorphism, the image is a proper subalgebra of M that is isomorphic to M. Such a case does not occur for finite dimensional algebras, of course, but this easily happens for infinite dimensional algebras.) We have a product operation here, defined as a composition of endomorphisms. We also have a direct sum operation as follows. (Note that if we just add two endomorphisms as maps, then the resulting map is not an endomorphism in general.) For two endomorphisms ρ, σ of M, we define a homomorphism $x \mapsto \begin{pmatrix} \rho(x) & 0 \\ 0 & \sigma(x) \end{pmatrix}$ from M to $M \otimes M_2(\mathbf{C})$. The image is in $M \otimes M_2(\mathbf{C})$ and not in M, but if M falls in a class of type III factor that naturally arises in the setting of algebraic quantum field theory that I will explain below, then we have an easy isomorphism from $M \otimes M_2(\mathbf{C})$ onto M, and using this, we obtain an endomorphism of M that is the direct sum of ρ, σ. Furthermore, we can define a notion of a dimension for endomorphisms, and it takes a value in $[1, \infty]$. This is essentially the same as the square root of the Jones index mentioned above. We also have a notion of irreducibility for endomorphisms, and if a dimension is finite, we can make an irreducible decomposition of an endomorphism in a unique way in an appropriate sense. The dimension is additive and multiplicative with respect to a direct sum and a composition, respectively, and an algebraic system of endomorphisms behaves similarly to an algebraic system of unitary representations of a compact group, where we have a direct sum and a tensor product. For an endomorphism, we have a notion of a conjugate endomorphism which is an analogue of a contragredient representation, and we have a direct analogue of the Frobenius reciprocity for conjugate endomorphisms. We have two differences between our endomorphisms here and unitary representations of compact groups, as follows. First, our dimensions are now not necessarily integers. Second, our compositions of endomorphisms are not necessarily commutative, though the tensor product operation, which is the corresponding operation in representation theory, is commutative in an obvious sense. (We have compositions of endomorphisms of an infinite dimensional algebra. There is no reason for such an operation to be commutative.) This algebraic system is a certain tensor category having endomorphisms as objects. We have a notion of amenability for such a category, and a tensor category having only finitely many irreducible objects is amenable, as an analogue of a finite group. In such a case, we say that a category is rational. The rules of irreducible decompositions of compositions of two irreducible objects are called the fusion rules. By comparing two kinds of irreducible decompositions of a product of three objects, we obtain $6j$-symbols, as in the representation theory of compact groups. We simply call them *$6j$-symbols*, or *quantum $6j$-symbols*. A paragroup is an abstract algebraic system and axiomatized with several properties of such categories of endomorphisms, like those on fusion rules and quantum $6j$-symbols. Popa's classification theorem says that if we have amenability for both factors and paragroups, then subfactors are completely classified by paragroups. We omit details on these categories, quantum $6j$-symbols, and paragroups, and refer the reader to the book [**9**] for precise definitions. Here we only make brief remarks, in comparison to similar structures appearing in conformal field theory and quantum groups, that the fusion rules here are not necessarily commutative and that the (quantum) dimensions are always positive now. Here we also remark that if a tensor category arising from a subfactor as above has only finitely many irreducible objects, then we can construct a complex-number-valued topological invariant of 3-dimensional oriented closed manifolds, and more generally, 3-dimensional topological quantum field theory, from such a category. This is a construction of Ocneanu, generalizing the one by Turaev and Viro. (Note that a category here does not have a braiding in general. We also have operator algebraic methods to produce a braiding.)

Now Popa's classification theorem mentioned above falls under the general principle, explained above, that an algebraic invariant becomes complete under amenability, and in this sense, the result itself is similar to other classification theorems, but we now have a new feature, as follows. A very nice abstract characterization of paragroups has been obtained by Ocneanu, but it is not clear at all what examples of paragroups we indeed have, and this problem can be studied within a framework of operator algebras. On one hand, finitely generated abelian groups and ergodic flows, which appear as invariants of operator algebraic objects, have been studied well for a long time. On the other hand, the history of the studies of tensor categories is relatively short, and not much is known beyond representation theories of groups and quantum groups having deformation parameters. We, however, think, from the viewpoint of subfactor theory, that tensor categories arising from quantum groups with a deformation parameter q are special; we have many other tensor categories, and operator algebras are useful for studying them. (The biggest evidence for such an idea is a combinatorial study of Haagerup [**12, 1**] for subfactors with small Jones index. He made an exhaustive search for paragroups in a very limited index range, found a large list of candidates that do not seem to be related to usual quantum group theory, and proved that two of them are indeed realized, after very hard computations.) So, even if one is interested only in tensor categories and topological invariants arising from them, not in operator algebras themselves, one can still hope to obtain new results not known in other fields by operator algebraic mathods. One informal reason for this possibility is that infinite dimensional operator algebras are so "large" that one can put many mathematical objects into them. In the case of operator algebraic invariants such as finitely generated abelian groups and ergodic flows, we do not obtain really new examples through operator algebras, so the new situation for tensor categories is different from these predecessors and interesting for this reason. This is a basis for a series of new relations between subfactor theory and "quantum something".

4. Algebraic quantum field theory

There have been different mathematical approaches to quantum field theory, and here I explain the one based on operator algebras. This approach relies on algebras of bounded linear operators, rather than operator valued distributions, and is called "algebraic" quantum field theory in this sense. A standard textbook is [**11**], and one can look at the proceedings volume [**18**] for recent developments. Here I present a basic setting below.

We fix a "space-time" and consider appropriate bounded regions in it, those called double cones for example. To each such region $\mathcal{O}$ we assign a von Neumann algebra $A(\mathcal{O})$ generated by physical observables in $\mathcal{O}$. (We are now in the realm of quantum mechanics, so physical quantities are represented as operators on a Hilbert space.) We have a fixed Hilbert space for all $\mathcal{O}$'s. Mathematically speaking, this gives a family of operator algebras parameterized by $\mathcal{O}$'s. We require a set of physically natural axioms to hold for such a family of operator algebras, and we study such families mathematically. Here I explain only the fundamental axioms, though we often assume various sets of extra axioms in concrete studies. First, we assume that we have a larger operator algebra for a larger space-time region $\mathcal{O}$, which means that we have more observable quantities in a larger region. Second, if two regions $\mathcal{O}_1$, $\mathcal{O}_2$ are spacelike and separated, then the two operator algebras $A(\mathcal{O}_1)$, $A(\mathcal{O}_2)$ commute, which is called *locality*. This means a relativistic requirement that if one cannot reach $\mathcal{O}_1$ from $\mathcal{O}_2$ even at the speed of light, then we have no interactions between them, and thus physical quantities in these two regions commute. Next we assume that we have a projective unitary representation u_g of a group of "space-time symmetry" on the Hilbert space, and the equality $u_g A(\mathcal{O}) u_g^* = A(g\mathcal{O})$ holds, which is called *covariance*. A physically natural choice of the space-time symmetry is the inhomogeneous Lorentz group of the 4-dimensional Minkowski space, but we will also consider other choices of the symmetry below. We also have axioms on a vacuum vector and positivity of energy, but I omit explanations of them here. We often call a family $\{A(\mathcal{O})\}$ of operator algebras a *net* of operator algebras, because the index set of double cones $\mathcal{O}$ is directed with respect to inclusions.

Now we have the mathematical problem of studying structures of such nets of operator algebras under a suitable set of axioms, and even classifying them with a certain algebraic invariant. Our algebraic invariant is the representation theory of a net of operator algebras. Suppose we have a net $\{A(\mathcal{O})\}$ of operator algebras. These operator algebras act on a Hilbert space from the beginning by definition, but we also consider their representations on other Hilbert spaces. The original Hilbert space has a special vector called a *vacuum vector* by axiom, so the original representation on this Hilbert space is called a *vacuum representation*. We then consider a compatible family of representations of the algebras $\{A(\mathcal{O})\}$ with the covariance property on another Hilbert space, without requiring a vacuum vector there, and study unitary equivalence classes of such representations. We further have notions of a direct sum, irreducible decomposition, a tensor product, and a dimension for such representations. Note that we now have representations of a family of algebras, so the notion of tensor product is not obvious at all here. (In quantum group theory, they have a coproduct, but we do not have such an operation here.) Also, considering a dimension in the ordinary sense is meaningless since it is always infinite. These difficulties are overcome with the Doplicher-Haag-Roberts (DHR) theory [**8**]. That is, one first shows that any representation, up to unitary equivalence, is realized as an endomorphism of a net of operator algebras, based on the Haag duality, which is a stronger form of locality. (Note that an endomorphism of a net of operator algebras, appropriately defined, gives a new representation on the original Hilbert space. The DHR theory says that any representation can be put in this form after a (possible) change within the unitary equivalence class.) Then the operation called a "tensor product" is defined as a composition of two endomorphisms, and a dimension is also defined nicely. It is obvious that this situation is similar to that in subfactor theory explained in the previous section. Actually, the DHR theory is much older than Jones's subfactor theory, and the relations between the two were clarified by Longo [**17**]. In the previous section, we saw that the composition operation of endomorphisms is not commutative, but here, due to locality in the axioms, we have commutativity of composition, up to unitary equivalence. In this way, we obtain a tensor category of representations of a net of operator algebras.

As I mentioned above, the most natural choice for the space-time would be the 4-dimensional Minkowski space. Many deep results have been obtained for this setting, but in connection to "quantum something", we have interesting mathematical structures on space-time of lower dimensions, such as 1, as in [**10**], for example. The one-dimensional case is not a "space-time" any more, but any dimension is allowed in axiomatic mathematical studies, and for the case of one space dimension and one time dimension, which is quite interesting both mathematically and physically in the context of conformal field theory, a net of operator algebras on the 2-dimensional Minkowski space is decomposed into a tensor product of two nets of operator algebras on one-dimensional spaces in an appropriate sense, and such nets are called *chiral nets* and have been well studied. One might suspect that problems in lower dimensions, particularly in dimension one, are mathematically easier, but this is not the case. For nets of operator algebras on a one-dimensional space, one takes non-empty bounded open intervals as $\mathcal{O}$ in the above, but then a subtle and interesting problem, which does not occur in dimension three or higher, arises from the fact that its complement is not connected. For example, a composition of two endomorphisms arising from representations is commutative up to unitary equivalence, as mentioned above, and this unitary operator producing the unitary equivalence in the one-dimensional setting brings a braiding. In this way, the representation category becomes braided. In dimension 3 or higher, this braiding automatically becomes trivial, but many non-trivial examples of interesting braiding happen in the one-dimensional setting, and braided tensor categories corresponding to the Wess-Zumino-Witten models are realized. Also, the dimensions of representations are automatically integers in a higher dimensional space-time, but we have non-integer dimensions of representations for nets of operator algebras on a one-dimensional space, and they are counterparts of quantum dimensions in quantum group theory.

5. Tensor category, modular invariant, α-induction

I have finally come to my own research topics. The contents of this section describe what I have been studying after writing the book [**9**]. I refer the reader to the references for all technical matters and briefly mention the main results.

5.1. Completely rational AQFT. As explained above, a net of operator algebras on one-dimensional "space-time" produces a braided tensor category, but in connection with conformal field theory and topological quantum field theory, we are interested in tensor categories which have only finitely many irreducible objects and a non-degenerate braiding. The finiteness condition is usually called *rationality*. The non-degeneracy of a braiding is equivalent to invertibility of the S-matrix, and a braided tensor category with this non-degeneracy and rationality is called a *modular tensor category*. See [**25**] for details on this notion. Our main results in [**16**] are that we have a simple operator-algebraic sufficient condition for the rationality of the representation category, and that the braiding is automatically non-degenerate under this condition. We call this operator-algebraic condition *complete rationality*. By results in [**26**], [**28**], we know that the nets of operator algebras arising from loop groups of $SU(n)$ [**22**] are completely rational. Xu further studied completely rational nets of operator algebras in detail in [**29, 30, 31**]. Roughly speaking, complete rationality is preserved under various operations for nets of operator algebras.

5.2. α-induction and modular invariant. As mentioned above, the representation theory of nets of operator algebras is quite similar to that of compact groups. The method of α-induction gives a machinery in the representation theory of nets of operator algebras, analogous to induction and restriction in the ordinary representation theory. As an analogy of a group and its subgroup, we consider a net of operator algebras and a net of subalgebras, and we produce a representation of a larger net from one of a smaller net. We use the name "induction" here because we pass from smaller algebras to larger algebras, but in the setting of nets of operator algebras, a larger net has a "smaller symmetry", so this is more like a restriction operation in some sense. This construction of α-induction was first defined in Longo and Rehren [**19**], and many interesting properties and examples were found by Xu [**27**]. Böckenhauer and Evans [**2**] further studied this construction. Around the same time as Longo and Rehren, Ocneanu [**21**] made a combinatorial study on the *A*-*D*-*E* Dynkin diagrams from a quite different motivation. In [**3**], we have proved that both definitions, Longo-Rehren and Ocneanu, can be generalized to a situation where we have a braiding in the sense of Rehren [**23**], and we have a unified definition there. By combining both ideas, we have obtained several results in [**4, 5, 14**]. We would like to construct a representation of a larger net of operator algebras from a representation of a smaller net, and use a braiding in the α-induction method. A braiding always comes in a pair of an overcrossing and an undercrossing, and depending on these, we have two kinds of α-induction. We distinguish them by using the $\pm$ symbol. Actually, we cannot make a genuine representation of a larger net from one of a smaller net in general, and we obtain only a "quasi"-representation in some sense. If we take the intersections of the "quasi"-representations arising from α-induction with a positive braiding and those with a negative braiding, then it exactly gives the representations of the larger net. If the representation category of the smaller net is modular (as in the completely rational case, for example), then we have a finite dimensional unitary representation of $SL(2, \mathbf{Z})$, and we can produce a matrix with non-negative integer entries in the commutant of the range of this unitary representation by using positive and negative α-inductions. This was proved in [**3**] under a very general assumption, based on work of Ocneanu and of Böckenhauer and Evans [**2**]. Such matrices have been studied well in general under the name of "modular invariants", and several classification theorems, starting with [**6**], have been obtained. (See [**7**] for example.) Our results enable one to interpret and realize such classification results with operator-algebraic methods. There has also been much study of tensor categories arising from α-induction which do not have a braiding in general.

5.3. A central charge and classification of nets of factors. Recall that a basic idea for classification is that we should have simple complete algebraic invariants for the amenable case. For nets of operator algebras, each von Neumann algebra $A(\mathcal{O})$ assigned to a space-time region $\mathcal{O}$ is always isomorphic to the unique hyperfinite type III_1 factor, an amenable factor, under the standard set of axioms, so the isomorphism class of each algebra is useless for classification of nets. We expect that representation theory for completely rational nets on the one-dimensional space would give a complete invariant, but no results of this type have been obtained. It is also totally unclear which modular tensor category arises as a representation category of a net of operator algebras. All the constructions so far have been obtained depending on the case. Thus the general classification and construction problems are very far from a satisfactory solution, but we have recently obtained a first classification theorem in this direction in [**15**] as follows. We consider nets of operator algebras on the one-dimensional space, and we compactify the space to S^1. (This is just for simplicity of various descriptions. We could work on $\mathbf{R}$ without compactification if we want. We also take the orientation-preserving diffeomorphism group of S^1 as the group of "space-time" symmetries. (It is not mathematically clear how strong this assumption is, but we have many interesting examples satisfying this condition.) Then we can define a real number, called the "central charge", for a net of operator algebras through the representation theory of the Virasoro algebra. It is well-known that if this central charge c is less than one, then it can take only values in a discrete series, and our main results in [**15**] say that in such a case we have a complete classification and listing of nets of operator algebras. We fully use the theories of α-induction and modular invariants mentioned above. The classification result is described with pairs of the A-D-E Dynkin diagrams.

6. Future developments

Classification of nets of operator algebras still has many open problems, such as finding a complete invariant, characterizing possible invariants, and listing all the possible nets. We also have various problems on relations among extra axioms in addition to the standard set of axioms. Furthermore, we have limited (or no) knowledge on its relations to quantum groups, vertex operator algebras, the monster group, the moonshine, and topological quantum field theory. I finish this survey hoping for further developments in these and other yet-unknown directions.

References

1. M. Asaeda & U. Haagerup, *Exotic subfactors of finite depth with Jones indices* $(5+\sqrt{13})/2$ *and* $(5+\sqrt{17})/2$, Comm. Math. Phys. **202** (1999) 1–63.
2. J. Böckenhauer & D. E. Evans, *Modular invariants, graphs and α-induction for nets of subfactors I*, Comm. Math. Phys. **197** (1998) 361–386. II **200** (1999) 57–103. III **205** (1999) 183–228.
3. J. Böckenhauer, D. E. Evans & Y. Kawahigashi, *On α-induction, chiral projectors and modular invariants for subfactors*, Comm. Math. Phys. **208** (1999) 429–487.
4. J. Böckenhauer, D. E. Evans & Y. Kawahigashi, *Chiral structure of modular invariants for subfactors*, Comm. Math. Phys. **210** (2000) 733–784.
5. J. Böckenhauer, D. E. Evans & Y. Kawahigashi, *Longo-Rehren subfactors arising from α-induction*, Publ. Res. Inst. Math. Sci. **37** (2001) 1–35.
6. A. Cappelli, C. Itzykson & J.-B. Zuber, *The A-D-E classification of minimal and $A_1^{(1)}$ conformal invariant theories*, Comm. Math. Phys. **113** (1987) 1–26.
7. P. Di Francesco, P. Mathieu & D. Sénéchal, "Conformal Field Theory", Springer-Verlag, Berlin-Heidelberg-New York, 1996.
8. S. Doplicher, R. Haag & J. E. Roberts, *Local observables and particle statistics*, I. Comm. Math. Phys. **23**, 199–230 (1971); II. **35**, 49–85 (1974).

9. D. E. Evans & Y. Kawahigashi, "Quantum Symmetries on Operator Algebras", Oxford University Press, Oxford, 1998.
10. K. Fredenhagen, K.-H. Rehren & B. Schroer, *Superselection sectors with braid group statistics and exchange algebras*, I. Comm. Math. Phys. **125** (1989) 201–226; II. Rev. Math. Phys. **Special issue** (1992) 113–157.
11. R. Haag "Local Quantum Physics", Springer-Verlag, Berlin-Heidelberg-New York, 1996.
12. U. Haagerup, *Principal graphs of subfactors in the index range* $4 < 3 + \sqrt{2}$, in "Subfactors — Proceedings of the Taniguchi Symposium, Katata" (ed. H. Araki, et al.), World Scientific, 1994, 1–38.
13. V. F. R. Jones, *Index for subfactors*, Invent. Math. **72** (1983) 1–25.
14. Y. Kawahigashi, *Generalized Longo-Rehren subfactors and α-induction*, Comm. Math. Phys. **226** (2002) 269–287.
15. Y. Kawahigashi & R. Longo, *Classification of Local Conformal Nets. Case* $c < 1$, to appear in Ann. Math., math-ph/0201015.
16. Y. Kawahigashi, R. Longo & M. Müger, *Multi-interval subfactors and modularity of representations in conformal field theory*, Comm. Math. Phys. **219** (2001) 631–669.
17. R. Longo, *Index of subfactors and statistics of quantum fields*, I. Comm. Math. Phys. **126** (1989); II. 217–247 & **130** (1990) 285–309.
18. R. Longo (ed.), "Mathematical Physics in Mathematics and Physics", Fields Inst. Commun. **30**, Amer. Math. Soc., Providence, RI, 2001.
19. R. Longo & K.-H. Rehren, *Nets of subfactors*, Rev. Math. Phys. **7** (1995) 567–597.
20. A. Ocneanu, *Quantized group, string algebras and Galois theory for algebras*, in *Operator algebras and applications, Vol. 2 (Warwick, 1987)*, (ed. D. E. Evans and M. Takesaki), London Mathematical Society Lecture Note Series **36**, Cambridge University Press, Cambridge, 1988, 119–172.
21. A. Ocneanu, *Paths on Coxeter diagrams: From Platonic solids and singularities to minimal models and subfactors* (Notes recorded by S. Goto), Part 5 in *Lectures on operator theory*, (ed. B. V. Rajarama Bhat et al.), The Fields Institute Monographs, **13**, Amer. Math. Soc., Providence, RI, 1999, pp. 243–323.
22. A. Pressley & G. Segal, "Loop Groups", Oxford University Press, Oxford, 1986.
23. K.-H. Rehren, *Braid group statistics and their superselection rules*, in: "The Algebraic Theory of Superselection Sectors", D. Kastler ed., World Scientific, Singapore, 1990, 333–355.
24. M. Takesaki ,"Theory of Operator Algebras. I, II, III", Springer-Verlag, Berlin-Heidelberg-New York, 2002.
25. V. G. Turaev, "Quantum Invariants of Knots and 3-Manifolds", Walter de Gruyter, Berlin-New York, 1994.
26. A. Wassermann, *Operator algebras and conformal field theory III: Fusion of positive energy representations of $SU(N)$ using bounded operators*, Invent. Math. **133** (1998) 467–538.
27. F. Xu, *New braided endomorphisms from conformal inclusions*, Comm. Math. Phys. **192** (1998) 347–403.
28. F. Xu, *Jones-Wassermann subfactors for disconnected intervals*, Comm. Contemp. Math. **2** (2000) 307–347.
29. F. Xu, *Algebraic coset conformal field theories I*, Comm. Math. Phys. **211** (2000) 1–44.
30. F. Xu, *3-manifold invariants from cosets*, preprint 1999, math.GT/9907077.
31. F. Xu, *Algebraic orbifold conformal field theories*, Proc. Nat. Acad. Sci. U.S.A. **97** (2000) 14069–14073.

DEPARTMENT OF MATHEMATICAL SCIENCES, UNIVERSITY OF TOKYO, TOKYO, 153-8914, JAPAN
E-mail address: yasuyuki@ms.u-tokyo.ac.jp

Translated by YASUYUKI KAWAHIGASHI

Amer. Math. Soc. Transl.
(2) Vol. **215**, 2005

Brownian motion and value distribution theory of holomorphic maps and harmonic maps

Atsushi Atsuji

1. Picard's theorem and Brownian motion

Let Z be complex Brownian motion on $\mathbf{C}$, namely $Z_t = X_t + \sqrt{-1}Y_t$, where X and Y are independent real valued Brownian motions and f a holomorphic function on $\mathbf{C}$. It is well-known that for some complex Brownian motion $\tilde{Z}$

$$f(Z_t) = \tilde{Z}_{\rho_t}, \quad \rho_t = \int_0^t |f'(Z_t)|^2 ds.$$

This beautiful relationship between Brownian motion and holomorphic functions is called conformal invariance of Brownian motion, and it was discovered by P. Lévy, who was a pioneer of the theory of Brownian motion. This ensures some possibility of investigating complex analysis by Brownian motion.

B. Davis gave an elegant proof of Picard's little theorem using this relationship ([**16**], [**17**]).[1]

Let us look at the trajectory of Z in the figure below. (Note that Brownian motion on $\mathbf{R}^2$ does not hit a point. $\hat{0}$ stands for identification of the origin with a neighborhood of the origin.)

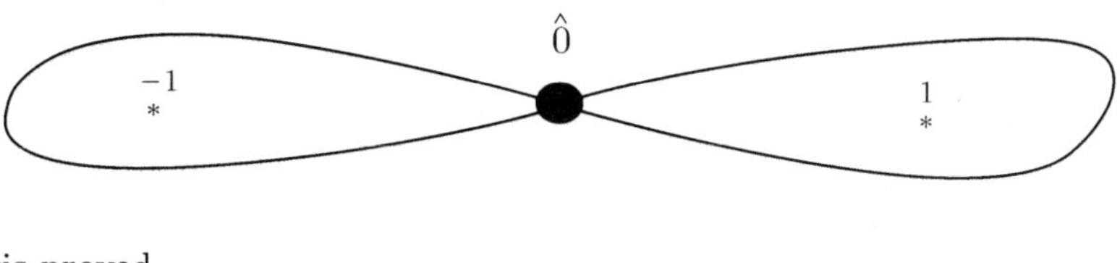

Davis proved

THEOREM 1. *Brownian motion on $\hat{\mathbf{C}} \setminus \{-1, 1\}$ starting from 0 makes a path homotopic to the one in the figure almost surely. Here $\hat{\mathbf{C}}$ is the complex plane which identifies the origin with $\hat{0}$.*

This article originally appeared in Japanese in Sūgaku **54** (3) (2002), 235–248.
2000 *Mathematics Subject Classification.* Primary 60J65; Secondary 30D35, 58J65.

[1]See also [**19**] and [**9**]. These books are good guides to the relationship between classical function theory and probability theory.

It is well-known that complex Brownian motion is recurrent. Namely, for any open subset $U \subset \mathbf{C}$,

$$\limsup_{t\to\infty} 1_U(Z_t) = 1, \quad \text{a.s.}^2$$

Hence Brownian motion starting from the origin makes a path coming back to a neighborhood of the origin $(\hat{0})$ with probability 1. Moreover, the strong Markov property of Brownian motion implies that it comes back to the neighborhood infinitely often. On the other hand, conformal invariance implies that a nonconstant holomorphic function f maps Brownian motion to a time-changed Brownian motion. Assume that f omits two points $\{-1, 1\}$. Theorem 1 implies that the image of Brownian motion by the holomorphic function makes a loop like the figure, while the Brownian motion on the domain of the function finitely often comes back to the neighborhood. Since $\mathbf{C}$ is simply connected and f is continuous, this leads us to a contradiction.

This proof is concerned with probabilistic and geometric properties of the domain and the target separately. This has for us the merit of making each role in the theorem clearer. The proof is a synthesis of them.

As we know, a universal covering of $\mathbf{C} \setminus \{-1, 1\}$ is conformally equivalent to the upper plane, which is a typical example of hyperbolic space. Equipped with the Poincaré metric, it becomes a complete space of constant negative curvature. By taking account of lifting it is natural to consider the case of simply connected and negatively curved manifolds as a target of maps. In fact, Davis used transience (i.e., non-recurrence) of a lifted process on the universal covering and the fact it converges to infinity as time tends to infinity in his proof.[3] If you extend these cases, keeping to the probabilistic nature of the proof, you will consider the situation that a process in the domain of maps has a property close to recurrence and the image process in the target converges to infinity as time tends to infinity. In this direction a probabilistic improvement of the Picard theorem was done by W. S. Kendall, who gave another version of the Goldberg-Ishihara-Petridis theorem ([**38**]).

2. Picard type theorems for harmonic maps and convergence of martingales

We exhibit a relationship between harmonic maps and martingales.

Consider a manifold with a connection. We can consider geodesics corresponding to the connection. We say that a function g on the manifold is geodesically convex if $t \to g(\gamma(t))$ is convex for any geodesic γ.

DEFINITION 2 (cf. [**22**]). Let $(\Omega, \mathcal{F}, \{\mathcal{F}_t\}, P)$ be a filtered probability space and Y a continuous stochastic process on $(\Omega, \mathcal{F}, \mathcal{F}_t, P)$ taking values in a manifold N with a connection. We say that Y is a Γ-martingale if $g(Y)$ is a submartingale on $(\Omega, \mathcal{F}, \{\mathcal{F}_t\}, P)$ for any convex function g defined locally on N.

A Γ-martingale on N is often called simply a martingale on N below. You can see that the notion of martingales is associated with a connection. "Γ-" comes from this.

---

[2] Probabilists have the habit of saying that an event occurs "a.s." (= "almost surely") instead of "with probability one".

[3] Davis proved Theorem 1 as a property of Brownian motion on $\mathbf{R}^2$. In this direction Goldberg and Mueller [**30**] generalized Davis' proof by studying the winding property of Brownian motion on Riemann surfaces to obtain a Picard theorem for meromorphic functions on Riemann surfaces.

We now give a definition of finely harmonic maps, which is a wider class than the usual harmonic maps and is related to Brownian motion, instead of the usual harmonic maps in the differential-geometric sense.

DEFINITION 3. A Borel measurable map f from a Riemannian manifold M to N is called a finely harmonic map if $h \circ f$ is subharmonic for any geodesically convex function h defined locally on N.

In particular, a smooth finely harmonic map is a usual harmonic map which is defined by a map of vanishing tension field (cf. [**58**]).

It would be interesting to know when a finely harmonic map becomes harmonic. This problem is also related to the Dirichlet problem of harmonic maps. For probabilistic approaches to these problems, see [**43**], [**3**], [**52**], [**15**].

The following shows a fundamental relationship between Brownian motion, martingales and harmonic maps.

THEOREM 4 (Bismut [**49**]). *A C^2-map $f : M \to N$ is harmonic if and only if f maps any Brownian motion on M to martingales on N.*

We next give a Picard type theorem for harmonic maps, due to Goldberg, Ishihara, and Petridis.

THEOREM 5 (Goldberg-Ishihara-Petridis [**29**]). *Let M be a complete Riemannian manifold of nonnegative Ricci curvature and N a Hadamard manifold (i.e. simply connected and of negative sectional curvature) whose sectional curvature is less than $-k < 0$. If $f : M \to N$ is a harmonic map of bounded dilatation, then f is constant.*

We say that $f : M \to N$ is of (K-)bounded dilatation if $\lambda^{(1)} \leq K^2\lambda^{(2)}$, where $\lambda^{(1)} \geq \lambda^{(2)} \geq \dots$ are eigenvalues of $df(df)^*$.

W. S. Kendall gave two types of proofs of this theorem.

THEOREM 6 (Kendall [**38**]). *Let M be a complete Riemannian manifold, N a Hadamard manifold whose sectional curvature satisfies $-a^2 \leq Sect_N \leq -b^2 < 0$. Let $f : M \to N$ be a harmonic map of bounded dilatation. If M admits no nonconstant bounded harmonic function, then f is constant.*

REMARK 7. If $Ric_M \geq 0$, M admits no nonconstant bounded harmonic function ([**60**]). Hence Theorem 6 requires a weaker assumption on the domains of maps than the Goldberg-Ishihara-Petridis theorem. But it needs a lower bound on the sectional curvature of the target manifold. Thus we cannot say which of them is the stronger result.

We consider probabilistic properties of M and N as in Davis' proof.

The assumption that M admits no nonconstant bounded harmonic function is equivalent to the triviality of an invariant σ-field of Brownian motion. Let Brownian motion be defined on a probability space $(\Omega, \mathcal{F}, P_x, x \in M)$.

An invariant σ-field is a collection of $A \in \mathcal{F}$ satisfying

$$1_A \circ \theta_t = 1_A, \quad t < \zeta, \quad P_x\text{-a.s.}, \quad \forall x \in M,$$

where ζ is the lifetime of the Brownian motion and $\theta_t : \Omega \to \Omega$ is defined by $\theta_t\omega(s) = \omega(t+s)$.

Let us extend the notion of "bounded dilatation" to martingales, in order to characterize martingales on N.

Let Y be a Γ-martingale on a d-dimensional Riemannian manifold N.

There exist a local martingale $\tilde{Y}$ on $\mathbf{R}^d$ and a random orthonormal frame $\Xi(t) : \mathbf{R}^d \to T_{Y(t)}$ satisfying a stochastic differential equation

$$d_S Y = \Xi d_S \tilde{Y},$$
$$d_S \Xi = H_\Xi d_S Y,$$

where $H_\xi : T_y N \to T_\xi O(N)$ ($O(N)$ is an orthonormal frame bundle) is the horizontal lift and d_S stands for the Stratonovich integral.[4] Ξ is called stochastic parallel transport and $\tilde{Y}$ is called stochastic development.

Set

$$\Lambda_t = \left(\frac{d\langle \tilde{Y}^i, \tilde{Y}^j \rangle_t}{dt} \right)$$

(assume its existence). We say Y is of (K-)bounded dilatation if the eigenvalues $\lambda^{(1)}(t) \geq \lambda^{(2)}(t) \geq \ldots$ of Λ_t satisfy $\lambda^{(1)} \leq K^2 \lambda^{(2)}$.

It is well-known that an n-dimensional Hadamard manifold with $-a^2 \leq Sect_N \leq -b^2 < 0$ has a geometric compactification $\overline{N}$ and its ideal boundary S_∞ is homeomorphic to the $n-1$-dimensional sphere S^{n-1}.

W. S. Kendall proved the following.

THEOREM 8. *Let Y be a nonconstant martingale of (K-)bounded dilatation on a Hadamard manifold N with $-a^2 \leq Sect_N \leq -b^2 < 0$. Then Y_t converges to a nontrivial random variable on S_∞ as $t \to \infty$ almost surely.*

We apply this to harmonic maps. Let $f : M \to N$ be a harmonic map of bounded dilatation. $Y_t = f(X_{t\wedge\zeta})$, where X is Brownian motion on M, is a martingale of bounded dilatation on N. $\lim_{t\to\infty} Y_t$ exists in S_∞ by the above theorem, but it is measurable for the invariant field of X. Then it should be constant. This leads to a contradiction.

3. A Picard type theorem of harmonic maps and coupling

We give another probabilistic proof of the Picard type theorem, using the coupling property. Coupling means that two processes starting from different points meet in finite time. Methods of coupling can be seen everywhere in probability theory (cf. [**46**]). As for applications to function theory, Lyons and Sullivan [**47**] used a relationship of existence of two Brownian motions for which coupling occurs with probability 1 and the Liouville property for bounded harmonic functions. Kendall used a slightly different coupling from theirs. Kendall required the two Brownian motions be adapted to the same filtration.

For simplicity we consider the case that M is stochastic complete (i.e., the lifetime of the Brownian motion is infinite a.s.).

DEFINITION 9. We say a Riemannian manifold M has BCP (Brownian coupling property) if there exist two Brownian motions X, Y ($X_0 \neq Y_0$), adapted to the same filtration, such that

$$P(X_t = Y_t \text{ for some } t,\ 0 < t < \infty) = 1.$$

[4]There are many notations for the Stratonovich integral, for instance, $\circ dY$ ([**35**]), δY ([**22**]), etc. Here we use Kendall's notation.

THEOREM 10 ([**39**]). 1) *Every complete Riemannian manifold with nonnegative Ricci curvature has BCP.*

2) *Let N be a Hadamard manifold with $Sect_N \leq -k < 0$. If X, Y are martingales of K-bounded dilatation with $X_0 \neq Y_0$, then coupling does not occur with positive probability, namely*

$$P(X_t = Y_t \text{ for some } t,\ 0 < t < \infty) < 1.$$

From this result the Goldberg-Ishihara-Petridis theorem can be immediately proved. Let $f : M \to N$ be a harmonic map of bounded dilatation, X and Y Brownian motions appearing in the definition of BCP. If f is nonconstant, then $f(X)$ and $f(Y)$ are martingales of bounded dilatation starting from distinct points in N. Though X and Y meet in finite time almost surely, $f(X)$ and $f(Y)$ never meet with positive probability. Hence f should be constant.

It seems that the function-theoretic and geometric meaning of BCP is still unclear. For instance,

QUESTION. Does a Riemannian manifold with BCP admit nonconstant positive harmonic functions?

4. Another type of Liouville theorem for harmonic maps

In this section we give a Liouville type theorem for harmonic maps whose domains are manifolds supporting Brownian motion with trivial invariant σ-fields.

THEOREM 11. *Let M be a Riemannian manifold supporting Brownian motion with trivial invariant σ-fields, N a Hadamard manifold and $f : M \to N$ a harmonic map.*

1) *If $f(M)$ is bounded in N, then f is constant.*

2) *If the energy of f is finite, namely,*

$$\int_M ||df||^2 dv_M < \infty,$$

then f is constant.

1) was shown by Kendall ([**42**]). 2) was by Cheng, Tam, and Wang ([**13**]). While ([**13**]) used an analytic method and reduced the problem to the first case, we here use a probabilistic method to show this. Since 1) and 2) can be seen similarly, we prove 2) here.

We first illuminate the meaning of finite energy from the martingale point of view.

DEFINITION 12. Let Y be a martingale on a Riemannian manifold (N, g_{ij}) and $\tilde{Y}$ the stochastic development of Y. An increasing process $[Y, Y]$ defined by

$$d[Y, Y]_t = \sum_i d\langle \tilde{Y}_i, \tilde{Y}_i \rangle_t$$

is called the intrinsic time of Y.

This is locally

$$d[Y, Y]_t = g_{ij}(Y) d\langle Y_i, Y_j \rangle_t.$$

This quantity is important to show the convergence of martingales. The following is known. Let $\overline{N}$ be the Alexandrov compactification of N.

THEOREM 13 (Darling [**14**]). *If* $[Y,Y]_\zeta < \infty$ *a.s., then* Y_t *converges in* $\overline{N}$ *as* $t \to \zeta$ *a.s.*

If f is of finite energy, X is transient, and we set $Y = f(X)$, we have, by Ito's formula,

$$E_o[[Y,Y]_\zeta] = \frac{1}{2}\int_M ||df||^2(x)g(o,x)dv(x),$$

where $X_0 = o$, dv is the Riemannian volume, and $g(x,y)$ is the 0-order Green function with respect to $\frac{1}{2}\Delta$. Hence

PROPOSITION 14. *Let* M, N *be Riemannian manifolds and* $f : M \to N$ *a harmonic map of finite energy. If Brownian motion on* M *is transient, then* f *converges in* $\overline{N}$ *along almost all Brownian paths.*

Let d be the distance on a Hadamard manifold N and $r(x) = d(o,x)$ for a reference point o. Since $r(x)$ is convex, $r \circ f$ is a subharmonic function. We also have

$$|\nabla r \circ f|^2(x) \le const.||df||^2(x).$$

Thus $r \circ f$ is a subharmonic function with finite Dirichlet integral. Now we consider the following two cases.

i) Brownian motion on M is transient.

ii) Brownian motion on M is recurrent.

i) Because of transience, and using the Royden decomposition ([**1**], [**2**], [**27**]) for a function on M in $L^2_{loc}(dv)$ of finite Dirichlet integral, we can see that if u is a nonnegative subharmonic function of finite Dirichlet integral, then u has a harmonic majorant h of finite Dirichlet integral. If X is Brownian motion, $u(X_t) \le h(X_t)$. Moreover, $h(X)$ is a local martingale and $\langle h(X)\rangle_t = \int_0^t |\nabla h|^2(X_s)ds$. By the Burkholder-Davis-Gundy inequality we have

$$E[\sup_{0t\le t\le\zeta} |h(X_t)|^2] \le C\,E[\int_0^\zeta |\nabla h|^2(X_s)ds] = C\int_M g(o,x)|\nabla h|^2(x)dv(x) < \infty,$$

for an arbitrary starting point $o \in M$. Hence $h(X)$ and $u(X)$ are uniformly integrable. Thus $\lim_{t\to\zeta} u(X_t)$ converges a.s. and in $L^1(dP)$.

On the other hand, the above proposition implies $\lim_{t\to\zeta} f(X_t)$ exists in N. This limit point is random in general, but is constant in this case since it is measurable for invariant σ-fields. Let y_o be this point and $u(x) = d(y_o, f(x))$. Since $u(X_t)$ is a uniformly integrable submartingale,

$$u(x) \le \lim_{t\to\infty} E_x[u(X_{t\wedge\zeta})] = 0.$$

Since $x \in M$ is arbitrary, we have $f(x) = y_o$.

ii) In the recurrent case, we have a stronger conclusion. We can deal with δ-subharmonic functions, which is a wider class than subharmonic functions. A δ-subharmonic function is a difference of two subharmonic functions locally. They play important roles in the value distribution theory of holomorphic maps. If v is a δ-subharmonic function, the corresponding Riesz measure ν is a signed measure which can be decomposed as $d\nu_v = d\nu_1 - d\nu_2$ (cf. [**32**], [**33**]), where $d\nu_1$ and $d\nu_2$ are nonnegative measures. When v is subharmonic, $d\nu_v$ is a nonnegative measure. If we take minimal $d\nu_1, d\nu_2$ for which the decomposition holds, this decomposition is unique.

From now on we deal only with such a unique decomposition. Let $d\tilde{\nu_v}, d\tilde{\nu_1}, d\tilde{\nu_2}$ be the smooth parts of $d\nu_v, d\nu_1, d\nu_2$ respectively; namely, they charge no set of zero capacity (cf. [**10**], [**28**]).

Our result is

THEOREM 15 ([**6**]). *Suppose M is a Riemannian manifold supporting recurrent Brownian motion and v is a δ-subharmonic function on M whose Riesz measure is $\nu_1 - \nu_2$ $(\nu_1, \nu_2 \geq 0)$. If v has a finite Dirichlet integral or is bounded, then v is constant or*

$$0 < \nu_1(M) = \tilde{\nu_1}(M) = \nu_2(M) = \tilde{\nu_2}(M) \ (\leq \infty).$$

REMARK 16. Takegoshi [**57**] proved a similar result for a class of smooth functions which are not always of finite Dirichlet integral, under a volume growth condition on M.

Ito's formula or the Fukushima decomposition for $v(X)$ ([**27**]) shows us that

$$v(X_t) - v(X_0) = B(\int_0^t |\nabla v|^2(X_s)ds) + A_t^{(1)} - A_t^{(2)}, \quad t < \zeta,$$

where $A^{(1)}, A^{(2)}$ are increasing additive functionals of X and their *Revuz* measures ([**27**]) are $d\tilde{\nu_1}$ and $d\tilde{\nu_2}$, respectively.

Now we note a ratio ergodic theorem following from recurrence.

LEMMA 17 (cf. [**54**]). *Let $C^{(1)}, C^{(2)}$ be positive continuous additive functionals of X. If the corresponding Revuz measures μ_1, μ_2 are finite,*

$$\lim_{t\to\infty} \frac{C_t^{(1)}}{C_t^{(2)}} = \frac{\mu_1(M)}{\mu_2(M)}, \quad a.s.$$

Assume that v is not harmonic and $\tilde{\nu_1}(M) > \tilde{\nu_2}(M)$. Then the ratio ergodic theorem implies that as $t \to \infty$, $v(X_t)$ is bounded below by a process like

$$V_t = B(\int_0^t |\nabla v|^2(X_s)ds) + const. \int_0^t |\nabla v|^2(X_s)ds.$$

The law of the iterated logarithm implies

$$\limsup_{t\to\infty} \frac{B_t}{\sqrt{2t \log\log t}} = 1, \quad \text{a.s.}$$

If $|\nabla v|$ is not identically zero, recurrence implies $\int_0^\infty |\nabla v|^2(X_s)ds = \infty$, a.s. Using them, we can see that $V_t \to \infty$ $(t \to \infty)$, a.s. Hence $v(X_t) \to \infty$, a.s. This contradicts the recurrence.

Thus $|\nabla v| = 0$; namely, v is constant. When v is harmonic, a similar procedure works well.

We apply this to a harmonic map and distance function. While in the previous results the target manifolds N are Hadamard manifolds which are free of cut locus at every point, we wish to consider the case that N may have cut loci. It is known that $r(x) = d(o, x)$ is a δ-subharmonic function, and for it the negative part of the Riesz measure is supported by cut locus of o ([**59**]). If $M \to N$ is a finely harmonic map, $r \circ f$ is also a δ-subharmonic function (cf. [**44**], [**8**]).

COROLLARY 18. *Suppose that M, N are Riemannian manifolds, M supports recurrent Brownian motion and $f : M \to N$ is a harmonic map. If f is of finite* *energy or its image is included in a compact set in N, then f is constant or, for any $o \in N$, $r(x) = d(o,x)$ and f satisfy*

$$0 < \tilde{\mu_1}(M) = \tilde{\mu_2}(M) \ (\leq \infty),$$

where $\mu_1 - \mu_2$ is the Riesz measure of $r \circ f$.

Theorem 6 and the results of this section can be extended to the case of L-finely harmonic maps. We give a definition of it. Suppose that L on a manifold M is a self-adjoint operator on some $L^2(M,m)$ and is associated with a local Dirichlet form $\mathcal{E}$. Thus L is a generator of a diffusion process X on M. We say that a Borel measurable map $f : M \to N$ is an L-finely harmonic map if $c \circ f$ is L-subharmonic for any convex function c defined locally on N. This is equivalent to the fact that $f(X)$ is a martingale on N. As for the energy of such maps, we can define it from the Revuz measure of the intrinsic time $[f(X), f(X)]$, since $[f(X), f(X)]$ is a positive continuous additive functional of X. Some relationship between such energy integrals and the related variational problem can be discussed (see [**53**]).

5. Nevanlinna theory and stochastic calculus

Nevanlinna theory is known as a quantitative refinement of Picard's theorem. We wish to note a probabilistic approach to Nevanlinna theory. While for harmonic maps we used qualitative properties like the recurrence of Brownian motion and the convergence theorem of martingales, we need some estimates and inequalities for probabilistic quantities for Nevanlinna theory. Let us recall the conformal invariance of Brownian motion. Information on holomorphic functions along a Brownian path can be divided into two parts: Brownian motion on the target and the new clock ρ_t of time-change. ρ_t is an additive functional of Brownian motion on the domain of maps. Thus it is important to consider inequalities for such functionals.

We first recall the classical case as follows. Let $f : \mathbf{C} \to \mathbf{C} \cup \{\infty\}$ be a meromorphic function. We define the following functions (cf. [**31**], [**34**]). Assume f is non-constant and $f(o) \neq a, \infty$. Then we set

$$m(a,r) = \int_0^{2\pi} \log[f(re^{i\theta}), a]^{-2} \frac{d\theta}{2\pi} \qquad \text{(proximity function)},$$

$$N(a,r) = \sum_{f(\zeta)=a, |\zeta|<r} \log \frac{r}{|\zeta|} \qquad \text{(counting function)},$$

$$T(r) = \int_{|z|<r} \frac{|f'(z)|^2}{(1+|f(z)|^2)^2} g_r(o,z) dv(z) \qquad \text{(Ahlfors-Shimizu characteristic function)},$$

where

$$[w,a] = \frac{|w-a|}{\sqrt{|w|^2+1}\sqrt{|a|^2+1}} \quad (\text{if } a \neq \infty), \qquad = \frac{1}{\sqrt{|w|^2+1}} \quad (\text{if } a = \infty),$$

$g_r(x,y)$ is the Green function with respect to $\frac{1}{2}\Delta$ on $|z| < r$ with Dirichlet boundary condition, and dv is Lebesgue measure. We know two fundamental theorems.

FIRST MAIN THEOREM. *For any $a \in \mathbf{C} \cup \{\infty\}$ and $0 < r < \infty$,*

$$m(a,r) + N(a,r) = T(r) + O(1).$$

SECOND MAIN THEOREM. *Let $a_1, a_2, \ldots, a_q \in \mathbf{C} \cup \{\infty\}$ be distinct points. There exists $E \subset [0, \infty)$ with $|E| < \infty$ such that*

$$\sum_{k=1}^{q} m(a_i, r) + N_1(r) \leq 2T(r) + O(\log T(r) + \log r)$$

for $r \notin E$.

If f is not constant, it is known that $T(r) \nearrow \infty$ ($r \nearrow \infty$) and $O(\log T(r) + \log r)$ in the above inequality can be reduced to $o(T(r))$. Hence we have Picard's little theorem from the above two theorems. We are going to see their probabilistic counterpart.

5.1. Stochastic first main theorem. Though we can get several interpretations of Nevanlinna's first main theorem, we can summarize them as follows.

By the Doob-Meyer decomposition a local submartingale U can be written as

$$U = M + A,$$

where M is a local martingale and A is an increasing process.

We get the following immediately from this.

THEOREM 19. *Let U be a nonnegative local submartingale and T a stopping time. Then*

$$E[U_T] - E[U_0] + N(T, U) = E[A_T],$$

where A is an increasing part of the Doob-Meyer decomposition of U and

$$N(T, U) = \lim_{\lambda \to \infty} \lambda P(\sup_{0 \leq t \leq T} U_t > \lambda).$$

If $N(T, U) = 0$, U is a pure submartingale. We say that a local martingale is a strictly local martingale if it is not a martingale. A martingale which is not strictly local is sometimes called a pure martingale. See [**20**], [**21**], [**55**] for properties and some related results.

Examples: From the stochastic first main theorem to the first main theorem in Nevanlinna theory.

i) Classical case 1. Let f be a meromorphic function on $\mathbf{C}$. We note that

$$f(Z_t) = \tilde{Z}_{\rho_t}, \quad \rho_t = \int_0^t |f'(Z_t)|^2 ds,$$

where $Z, \tilde{Z}$ are complex Brownian motions. Set $u_a(w) = \log[w, a]^{-1}$. u_a is a δ-subharmonic function. $U = u_a(\tilde{Z})$ is a local submartingale, since the negative part of the Riesz measure of $u_a \circ f$ is supported by a set of zero capacity.

Set $T = \rho_{\tau_r}$, $\tau_r = \{t > 0 : |Z_t| > r\}$. We have

$$E[A_T] = T(r), \quad N(T, U) = N(a, r), \quad E[U_T] = m(a, r).$$

ii) Classical case 2. It is rather natural to regard the image of complex Brownian motion by a meromorphic function as a stochastic process on 1-dimensional complex projective space $\mathbf{P}^1(\mathbf{C})$. We call a diffusion process associated with a half of the Laplacian defined from the Fubini-Study metric on $\mathbf{P}^1(\mathbf{C})$ a Brownian motion on $\mathbf{P}^1(\mathbf{C})$, denoted here by W. On $\mathbf{C}$ we have

$$W_{\eta_t} = Z_t, \quad \eta_t = \int_0^t \frac{ds}{(1 + |Z_s|^2)^2},$$

for a complex Brownian motion Z. If $f : \mathbf{C} \to \mathbf{P}^1(\mathbf{C})$ is a holomorphic map, then

$$\tilde{W}_{\tilde{\eta}_t} = f(Z_t), \quad \tilde{\eta}_t = \int_0^t \frac{|f'(Z_s)|^2}{(1+|f(Z_s)|^2)^2} ds,$$

for some Brownian motion $\tilde{W}$ on $\mathbf{P}^1(\mathbf{C})$. Setting $U = u_a(W)$ and $T = \tilde{\eta}_{\tau_r}$, we have the classical first main theorem. This relationship was found by Carne ([**12**]).

5.2. A probabilistic meaning of characteristic functions. For a generalization of the above observation we introduce a holomorphic martingale as image processes by holomorphic maps. For basic properties and applications, the reader can be referred to H. Kaneko's old article in Sugaku ([**37**]).

DEFINITION 20 ([**25**], [**37**]). Let N be a complex manifold and h any holomorphic function defined locally on N. A continuous stochastic process Y on N is a holomorphic martingale if $\operatorname{Re} h(Y)$ is a local martingale. A diffusion process which is a holomorphic martingale is called a holomorphic diffusion.

Brownian motion associated with a Kähler metric on a Kähler manifold is a holomorphic diffusion. The following properties are well-known ([**26**], [**36**]).

PROPOSITION 21. i) *If $f : M \to N$ is a holomorphic map between complex manifolds M and N and X is a holomorphic martingale on M, then $f(X)$ is a holomorphic martingale on N.*

ii) *If X is a holomorphic martingale on M and u is a plurisubharmonic function on M, then $u(X)$ is a local submartingale.*

If $(N, (g_{\alpha,\overline{\beta}}))$ is a Kähler manifold and Y is a holomorphic martingale on N, Y is a Γ-martingale. The intrinsic time $[Y, Y]$ of Y is defined by

$$d[Y,Y]_t = g_{\alpha,\overline{\beta}}\, d\langle Y^\alpha, \overline{Y}^\beta\rangle_t.$$

Let M be a Kähler manifold and N a compact Kähler manifold with a positive line bundle L. For a nonconstant holomorphic map $f : M \to N$ we define a characteristic function of f by

$$T(r) = E[[f(X), f(X)]_{\tau_r}],$$

where the intrinsic time of $f(X)$ is defined from the curvature form of L and $\tau_r = \inf\{t > 0 : X_t \notin D_r\}$ with an exhaustion $\{D_r\}$ of M.

By Darling's martingale convergence theorem and the stochastic first main theorem we have

THEOREM 22 ([**5**]). *If M admits no nonconstant bounded harmonic function and f is a nonconstant holomorphic map, then $T(r) \nearrow \infty$ as $r \to \infty$.*

From this we immediately have a generalization of the Casorati-Weierstrass theorem.

COROLLARY 23. *If f is a nonconstant holomorphic map from a Kähler manifold which admits no nonconstant bounded harmonic function to complex projective space $\mathbf{P}^n(\mathbf{C})$, then f can omit no sets of hyperplanes of positive measure in $\mathbf{P}^n(\mathbf{C})$.*

REMARK 24. "Positive measure" in the above statement can be replaced by "positive projective capacity". Projective capacity was introduced by Molzon [**50**].

Any algebraic submanifold in $\mathbf{C}^n$ admits no nonconstant bounded harmonic function ([**56**]). Hence

COROLLARY 25. *The Casorati-Weierstrass theorem holds on any algebraic submanifold in* $\mathbf{C}^n$.

5.3. Second main theorem. In this section we exhibit a probabilistic approach to the second main theorem. The following two lemmas are essential.

LEMMA 26 ([**4**]). i) *Let Y a Brownian motion on a two-dimensional compact Riemannian manifold M, and let $k \in L^{1+\epsilon}(M, dv)$ $(\epsilon > 0)$, $k \geq 0$. There exist constants $c_1, c_2 > 0$ such that, for any stopping time T,*

$$E[\int_0^T k(Y_t)dt] \leq c_1 E[T] + c_2.$$

ii) *Let Y be a martingale on a d-dimensional compact Riemannian manifold M, and let $k \in L^d(M, dv)$, $k \geq 0$. There exist constants $C_1, C_2 > 0$ such that for any stopping time T*

$$E[\int_0^T k(Y_t)\xi_t dt] \leq C_1 E[[Y,Y]_T] + C_2,$$

where

$$\xi_t = \det\left(\left(\frac{d\langle \tilde{Y}_i, \tilde{Y}_j\rangle_t}{dt}\right)\right)^{1/d}$$

and $\tilde{Y}$ is the stochastic development of Y.

LEMMA 27 ([**4**]). *Let X be Brownian motion on $\mathbf{R}^d$ and $K(x)$ a nonnegative locally integrable function on $\mathbf{R}^d$. For $\delta > 0$ there exists $E_\delta \subset [0, \infty)$ with $|E_\delta| < \infty$ such that for $r \notin E_\delta$*

$$E[K(X_{\tau_r})] \leq C(r)(E[\int_0^{\tau_r} K(X_t)dt])^{(1+\delta)^2}.$$

Here $C(r) = const.r^{(d-1)\delta}$.

REMARK 28. Lemma 27 holds in the cases of Hadamard manifolds with bounded sectional curvature, parabolic Riemann surfaces, and some class of minimal submanifolds in $\mathbf{R}^d$ ([**4**]). A key to the proof is a comparison theorem for stochastic differential equations ([**35**]) and comparison of Green functions derived from it ([**18**], [**48**]).

It is easy to derive the second main theorem in some cases from these inequalities. In particular, using Lemmas 26 and 27, we have another version of the second main theorem for meromorphic functions on a class of manifolds. Let k be a function on $\mathbf{P}^1(\mathbf{C})$ satisfying the assumption of Lemma 26 i), and M a complex manifold for which the inequality in Lemma 27 holds for some $C(r)$.

Let $f : M \to \mathbf{P}^1(\mathbf{C})$ be a nonconstant meromorphic function on M. Set $K = k \circ f \frac{f^*\omega}{|dz|^2}$ with the Fubini-Study form ω on $\mathbf{P}^1(\mathbf{C})$.

By time-change we have

$$\int_0^{\tau_r} K(X_t)dt = \int_0^{\eta_{\tau_r}} k(W_t)dt.$$

By Jensen's inequality,

$$\begin{aligned}E[\log K(X_{\tau_r})] &\le const. \log E[\int_0^{\tau_r} K(X_t)dt] + \log C(r) + O(1) \text{ (by Lemma 27)}\\ &= const. \log E[\int_0^{\eta_{\tau_r}} k(W_t)dt] + \log C(r) + O(1)\\ &\le const. \log E[\eta_{\tau_r}] + \log C(r) + O(1) \text{ (by Lemma 26 i))}\\ &= const. \log T(r) + \log C(r) + O(1).\end{aligned}$$

On the other hand, the first main theorem implies that

$$E[\log K(X_{\tau_r})] = -2T(r) + N_1(r) + E[\log k \circ f(X_{\tau_r})] + O(1).$$

Hence we have a version of the second main theorem, since

$$E[\log k \circ f(X_{\tau_r})] + N_1(r) \le 2T(r) + O(\log T(r)) + \log C(r) + O(1)$$

for r outside an exceptional set of finite length.

As for the usual type of second main theorem, we note that the inequality in Lemma 26 holds for a class of functions on $\mathbf{P}^1(\mathbf{C})$ introduced by Griffiths ([**31**], [**45**]), namely,

LEMMA 29. *Let*

$$k(w) = \frac{1}{\prod_{j=1}^q [w, a_j]^2 (\log[w, a_j])^2},$$

for distinct points $a_1, a_2, \dots, a_q$ and a Brownian motion Y on $\mathbf{P}^1(\mathbf{C})$. Then there exist constants $c_1, c_2 > 0$ such that, for any stopping time T,

$$E[\int_0^T k(Y_t)dt] \le c_1 E[T] + c_2.$$

Set $K(z)dzd\bar{z} = k \circ f\, f^*\omega$. Then, by the first main theorem and Jensen's inequality,

$$\begin{aligned}\sum_{j=1}^q m(a_j, r) + N_1(r) &= E[\log K(X_{\tau_r})] + \sum_{j=1}^q E[\log(\log[f(X_{\tau_r}), a_j])^2]\\ &\le E[\log K(X_{\tau_r})] + O(\log T(r)).\end{aligned}$$

The last term can be estimated as before. Hence we have

$$\sum_{j=1}^q m(a_j, r) + N_1(r) \le 2T(r) + O(\log T(r)) + \log C(r) + O(1)$$

for r outside an exceptional set of finite length.

It is a problem whether or not $\log C(r) = o(T(r))$. But it is not simple except in the case of $M = \mathbf{C}^n$. For example, in the case of Hadamard manifolds with sectional curvature pinched by negative constants we have $\log C(r) = O(r)$, and there exists a bounded holomorphic function which satisfies $T(r) = O(1)$. It may be interesting to consider estimates on $C(r)$ when a manifold admits no constant bounded harmonic functions. It may also be interesting to know any probabilistic meaning of this 'remainder term'.

Using Lemma 26 ii) enables us to treat the case of higher dimensional targets. But Lemma 26 ii) works only in the equidimensional case. Instead of this demerit we can relax some part of the assumptions of nonsingularity and normal crossing of divisors of the target manifolds in the results of Carlson and Griffiths [**11**] by Lemma 26 ii). It would be interesting to see if such an inequality as Lemma 26 ii) holds for general martingales.

Eremenko and Sodin ([**23**], [**24**]) obtained a second main theorem in non-equidimensional cases, in particular a generalization of the second main theorem for holomorphic curves in $\mathbf{P}^n(\mathbf{C})$ due to H. Cartan (cf. [**51**]), by using potential theory on the complex plane. It seems closely related to the probabilistic nature of the second main theorem, but the author has no idea about it yet.

Nevanlinna's original proof of his second main theorem was based on his *lemma on the logarithmic derivative*. We apply our technique to this. We can generalize this to an estimate for a class of δ-subharmonic functions ([**6**]). Such an estimate has applications outside complex analysis. For instance, we can see that if a minimal surface of finite genus and with finite number of ends has finite projective volume, then it has finite total curvature, even if it is not properly immersed.

Acknowledgement

The author would like to thank the referee for a careful reading of this article and helpful advice.

References

[1] A. Ancona, Théorie du potentiel sur les graphes et les variétés. École d'été de Probabilités de Saint-Flour XVIII—1988, 1-112, Lecture Notes in Math., 1427, Springer, Berlin, 1990.

[2] A. Ancona, R. Lyons, Y. Peres, Crossing estimates and convergence of Dirichlet functions along random walk and diffusion paths. Ann. Probab. 27 (1999), no. 2, 970-989.

[3] M. Arnaudon, X.-M. Li, A. Thalmaier, Manifold-valued martingales, changes of probabilities, and smoothness of finely harmonic maps. Ann. Inst. H. Poincaré Probab. Statist. 35 (1999), no. 6, 765-791.

[4] A. Atsuji, Nevanlinna theory via stochastic calculus. J. Funct. Anal. 132 (1995), no. 2, 473–510.

[5] A. Atsuji, A Casorati-Weierstrass theorem for holomorphic maps and invariant σ-fields of homomorphic diffusions. Bull. Sci. Math. 123 (1999), no. 5, 371-383.

[6] A. Atsuji, A lemma of logarithmic derivative for some δ-subharmonic functions. Complex Variables 46 (2001), no. 3, 195-206.

[7] A. Atsuji, Parabolicity, divergence theorem for δ-subharmonic functions and its application to Liouville theorems for harmonic maps, preprint.

[8] D. Barden, H. Le, Some consequences of the nature of the distance function on the cut locus in a Riemannian manifold. J. London Math. Soc. (2) 56 (1997), 369-383

[9] R. Bass, *Probabilistic techniques in analysis.* Springer, New York, 1995.

[10] R. M. Blumenthal, R. K. Getoor, *Markov processes and potential theory.* Pure and Applied Mathematics, Vol. 29, Academic Press, New York-London, 1968.

[11] J. Carlson, P. Griffiths, A defect relation for equidimensional holomorphic mappings between algebraic varieties. Ann. of Math. (2) 95 (1972), 557–584.

[12] T. K. Carne, Brownian motion and Nevanlinna theory. Proc. London Math. Soc. (3) 52 (1986), no. 2, 349–368.

[13] S. Y. Cheng, L.-F. Tam, T. Y.-H. Wan, Harmonic maps with finite total energy. Proc. Amer. Math. Soc. 124 (1996), no. 1, 275-284.

[14] R. W. R. Darling, Convergence of martingales on a Riemannian manifold. Publ. Res. Inst. Math. Sci. 19 (1983), no. 2, 753–763.

[15] R. W. R. Darling, Martingales on noncompact manifolds: maximal inequalities and prescribed limits. Ann. Inst. H. Poincaré Probab. Statist. 32 (1996), no. 4, 431-454.

[16] B. Davis, Picard's theorem and Brownian motion. Trans. Amer. Math. Soc. 213 (1975), 353-362.

[17] B. Davis, Brownian motion and analytic functions. Ann. Probab. 7 (1979), no. 6, 913-932.

[18] A. Debiard, B. Gaveau, E. Mazet, Théorèmes de comparaison en géométrie riemannienne. Publ. Res. Inst. Math. Sci. 12 (1976/77), no. 2, 391–425.

[19] R. Durrett, *Brownian motion and martingales in analysis*. Wadsworth, Belmont, CA, 1984.

[20] K. D. Elworthy, X.-M. Li, M. Yor, The importance of strictly local martingales; applications to radial Ornstein-Uhlenbeck processes. Probab. Theory Related Fields 115 (1999), no. 3, 325–355.

[21] K. D. Elworthy, X.-M. Li, M. Yor, On the tails of the supremum and the quadratic variation of strictly local martingales. Séminaire de Probabilités, XXXI, 113–125, Lecture Notes in Math., 1655, Springer, Berlin, 1997.

[22] M. Emery, *Stochastic calculus in manifolds*. Springer, Berlin, 1989.

[23] A. E. Eremenko, M. L. Sodin, Distribution of values of meromorphic functions and meromorphic curves from the standpoint of potential theory. St. Petersburg Math. J. 3 (1992), no. 1, 109-136.

[24] A. E. Eremenko, M. L. Sodin, The second fundamental theorem in value distribution theory of meromorphic curves for nonlinear divisors. Soviet Math. Dokl. 41 (1990), no. 2, 359-362 (1991).

[25] M. Fukushima, On holomorphic diffusions and plurisubharmonic functions. *Geometry of random motion*, 65-78, Contemp. Math., 73, Amer. Math. Soc., Providence, RI, 1988.

[26] M. Fukushima, M. Okada, On Dirichlet forms for plurisubharmonic functions. Acta Math. 159 (1987), no. 3-4, 171-213.

[27] M Fukushima, Y. Oshima, M. Tkeda, *Dirichlet forms and symmetric Markov processes*. de Gruyter Studies in Mathematics, 19, Walter de Gruyter & Co., Berlin, 1994.

[28] M. Fukushima, K. Sato, S. Taniguchi, On the closable parts of pre-Dirichlet forms and the fine supports of underlying measures. Osaka J. Math. 28 (1991), no. 3, 517-535.

[29] S. I. Goldberg, T. Ishihara, N. C. Petridis, Mappings of bounded dilatation of Riemannian manifolds. J. Differential Geometry 10 (1975), no. 4, 619–630.

[30] S. I. Goldberg, C. Mueller, Brownian motion, geometry, and generalizations of Picard's little theorem. Ann. Prob. 11 (1983), 833-846.

[31] P. A. Griffiths, *Entire holomorphic mappings in one and several complex variables*. Annals of Mathematics Studies, No. 85, Princeton University Press, Princeton, NJ, 1976.

[32] W. K. Hayman, P. B. Kennedy, *Subharmonic functions. Vol. I*. London Mathematical Society Monographs, No. 9, Academic Press, London-New York, 1976.

[33] W. K. Hayman, *Subharmonic functions. Vol. 2*. London Mathematical Society Monographs, 20, Academic Press, London, 1989.

[34] W. K. Hayman, *Meromorphic functions*, Oxford Mathematical Monographs, Clarendon Press, Oxford, 1964.

[35] N. Ikeda, S. Watanabe, *Stochastic differential equations and diffusion processes. Second edition.* North-Holland Mathematical Library, 24, North-Holland Publishing Co., Amsterdam; Kodansha, Ltd., Tokyo, 1989.

[36] H. Kaneko, A stochastic approach to a Liouville property for plurisubharmonic functions. J. Math. Soc. Japan 41 (1989), no. 2, 291–299.

[37] H. Kaneko, Plurisubharmonic functions and holomorphic diffusions on complex manifolds. Sugaku 41 (1989), 345–357 (Japanese).

[38] W. S. Kendall, Brownian motion and a generalised little Picard's theorem. Trans. Amer. Math. Soc. 275 (1983), no. 2, 751-760.

[39] W. S. Kendall, Stochastic differential geometry, a coupling property, and harmonic maps. J. London Math. Soc. (2) 33 (1986), no. 3, 554–566.

[40] W. S. Kendall, Nonnegative Ricci curvature and the Brownian coupling property. Stochastics 19 (1986), no. 1-2, 111–129.

[41] W. S. Kendall, Martingales on manifolds and harmonic maps. *Geometry of random motion*, 121–157, Contemp. Math., 73, Amer. Math. Soc., Providence, RI, 1988.

[42] W. S. Kendall, Probability, convexity, and harmonic maps with small image. I. Uniqueness and fine existence. Proc. London Math. Soc. (3) 61 (1990), no. 2, 371–406.

[43] W. S. Kendall, Convex geometry and nonconfluent Γ-martingales. I. Tightness and strict convexity. *Stochastic analysis* (Durham, 1990), 163–178, London Math. Soc. Lecture Note Ser., 167, Cambridge Univ. Press, Cambridge, 1991.

[44] W. S. Kendall, The radial part of a Γ-martingale and a non-implosion theorem. Ann. Prob. 23 (1995), no. 2, 479-500.

[45] S. Lang, W. Cherry, *Topics in Nevanlinna theory.* Lecture Notes in Mathematics, 1433, Springer-Verlag, Berlin, 1990.
[46] T. Lindvall, *Lectures on the coupling method.* Wiley Series in Probability and Mathematical Statistics, Wiley-Interscience, New York, 1992.
[47] T. Lyons, D. Sullivan, Function theory, random path, and covering spaces. J. Diff. Geom. 19 (1984), 299-323.
[48] P. Malliavin, Fonctions de Green d'un ouvert strictement pseudo-convexe et classe de Nevanlinna. C. R. Acad. Sci. Paris Ser. A 278 (1974), 141-144.
[49] P. A. Meyer, Géométrie stochastique sans larmes. Sém. Prob. XV, Springer Lecture Notes in Mathematics 850 (1981), 44-102.
[50] R. E. Molzon, Potential theory on complex projective space: application to characterization of pluripolar sets and growth of analytic varieties. Illinois J. Math. 28 (1984), no. 1, 103–119.
[51] J. Noguchi, T. Ochiai, *Geometric function theory in several complex variables.* Translations of Mathematical Monographs, 80, American Mathematical Society, Providence, RI, 1990.
[52] J. Picard, Smoothness of harmonic maps for hypoelliptic diffusions. Ann. Probab. 28 (2000), no. 2, 643–666.
[53] J. Picard, The manifold-valued Dirichlet problem for symmetric diffusions. Potential Anal. 14 (2001), no. 1, 53–72.
[54] D. Revuz, M. Yor, *Continuous martingales and Brownian motion.* Springer, Berlin, 1991.
[55] K. Takaoka, Some remarks on the uniform integrability of continuous martingales. Séminaire de Probabilités, XXXIII, 327–333, Lecture Notes in Math., 1709, Springer, Berlin, 1999.
[56] K. Takegoshi, Energy estimates and Liouville theorems for harmonic maps. Ann. Sci. École Norm. Sup. (4) 23 (1990), no. 4, 563–592.
[57] K. Takegoshi, A note on divergence of L^p-integrals of subharmonic functions and its applications. Proc. Amer. Math. Soc. 131 (2003), 2849–2858.
[58] H. Urakawa, *Calculus of variations and harmonic maps.* Translations of Mathematical Monographs, 132, American Mathematical Society, Providence, RI, 1993.
[59] S. T. Yau, Some function-theoretic properties of complete Riemannian manifolds and their applications to geometry. Indiana Univ. Math. J. 25 (1976), no. 7, 659-670.
[60] S. T. Yau, Harmonic functions on complete Riemannian manifolds. Comm. Pure Appl. Math. 28 (1975), 201-228.

DEPARTMENT OF MATHEMATICS, FACULTY OF ECONOMICS, KEIO UNIVERSITY, HIYOSHI, YOKOHAMA 223-8521, JAPAN

Translated by ATSUSHI ATSUJI

Amer. Math. Soc. Transl.
(2) Vol. **215**, 2005

Statistical inference in two–stage sampling

Makoto Aoshima

1. Introduction

Let $X_1, ..., X_n$ be i.i.d. random variables with a normal distribution $N(\mu, \sigma^2)$. Consider a hypothesis testing $H_0 : \mu = \mu_0$ against $H_1 : \mu = \mu_1$ $(\mu_0 \neq \mu_1)$. Stein [S45] proved a result originally given by Dantzig [D40] that there does not exist a critical region such that the probability of error of the first kind is less than the power for the testing of the mean when the scale parameter σ is unknown, as long as one considers a single–stage sampling scheme in which n is fixed in advance. That argument was extended to the case of the location–scale family of distributions by Chatterjee [C91]. Consider a fixed–width confidence interval for μ or a bounded risk point estimation for μ with the squared loss. For the location–scale family of distributions with unknown scale parameter, Lehmann [L51] proved that there does not exist such an estimation for the location parameter μ as long as one considers a single–stage sampling scheme. Both these conclusions essentially follow from the fact that two distributions with different locations can be made to approach each other uniformly in measure arbitrarily closely by making the scale parameter sufficiently large. Recently, Takada [T98] gave a sufficient condition for the nonexistence of such statistical procedures in terms of the distance of distributions under a single–stage sampling scheme.

To handle the inference described above, one is required to perform at least two stages of sampling. For most cases, two–stage sampling schemes successfully give a solution to those problems about fixed–size inference. There are cases, however, in which no sampling scheme exists, with a predetermined number of stages, that can yield a fixed–size inference. Then, purely sequential sampling schemes are required. (See Farrell [F66], Blum and Rosenblatt [BR66] and Koopmans et al. [KOR64].) Moreover, there are cases in which there does not exist any sampling scheme that can guarantee fixed–size inference. (See Bahadur and Savage [BaS56] and Blum and Rosenblatt [BR69a] for such examples. As for necessary and sufficient conditions about existence of certain multistage or purely sequential sampling schemes for fixed–size inference, refer to Blum and Rosenblatt [BR63] and Singh [Sin63].) Zacks [Z71, Chap.10] gives rich sources of this context. This article deals with inference problems which should be solved in at least two stages of sampling.

This article originally appeared in Japanese in Sūgaku **54** (4) (2000), 365–382.
2000 *Mathematics Subject Classification.* Primary 62D05.

The two–stage procedure is a statistical method originally created by Stein [S45] in 1945 to give a solution to inference problems, whose precision was fixed in advance, with the smallest number of required stages. In that procedure, the size of the second–stage sample is a random variable that is defined after observing the first–stage sample. The two–stage procedure has been applied to various fields of research and has had an impact on the development of the following methodologies: The purely sequential procedure given by Chow and Robbins [CR65] aims to reduce the average sample number asymptotically by taking a sample one by one repeatedly till it gets an approximate solution. It was also followed by the three–stage procedure given by Hall [Ha81], the accelerated sequential procedure given by Hall [Ha83], the modified accelerated sequential procedure given by Liu [Li97a], the multistage procedure given by Liu [Li97b] and the modified three–stage procedure given by Holm [Hol95, 99]. Even though development of methodology advances, the two–stage procedure does not lose its allure as an object of study. This is because of its utility. For instance, there is the case when one wants to make a statistical decision with demanded precision as soon as possible. There is the case when the cost of sampling becomes serious. There is the case when one fears that repeated observation will change the state of the population. For all these problems, which can happen in the real world, it may be said that the two–stage procedure is the easiest method to use.

The basic principle was invented by Stein in [S45]. The two–stage procedure is really elegant and marvelously lucid. Since then, many researchers have taken hints from this basic principle. In this article, without losing the essence of the idea, we shall consider later developments of the two–stage procedure mainly for bounded risk problems. For testing problems due to the two–stage procedure, one may refer to Hewett and Spurrier [HeS83].

2. Essence of Stein's two–stage procedure

We consider inference problems on the location parameter θ of a location–scale distribution $F((x-\theta)/\xi))$. The purpose of this section is to set forth the essence of Stein's two–stage procedure and give help in understanding the main subject of this article after Section 3.

Let $\{X_i;\ i \geq 1\}$ be i.i.d. with a p.d.f. $\xi^{-1}f((x-\theta)/\xi)$, where the unknown parameters $(\theta, \xi) \in R \times R^+$. Here, $f(\cdot)$ is known and continuous, and its domain space may depend on θ alone.

Let $T_n = T_n(X_1, ..., X_n)$ and $U_n = U_n(X_1, ..., X_n)$ be estimators of θ and ξ respectively based on samples $(X_1, ..., X_n)$. Suppose that T_n and U_n satisfy the following conditions:

(a) For any fixed n (≥ 2), T_n is independent of $(U_2, ..., U_n)$.

(b) (i) For some $\gamma > 0$, for a measurable function $g:\ R^+ \to R^+$, the distribution of $n^\gamma(T_n - \theta)/g(\xi)$ does not depend on (n, θ, ξ);

(ii) The distribution of $n^\gamma(T_n - \theta)/g(U_n)$ does not depend on (θ, ξ).

Given preassigned width $2d > 0$ and confidence coefficient $1 - \alpha > 0$, let us construct a confidence interval for θ by using Stein's two–stage procedure. Having recorded $X_1, ..., X_m$ of a suitable size m (≥ 2), calculate U_m and define

$$N = \max\{m,\ [(b_m g(U_m)/d)^{1/\gamma}] + 1\}, \tag{1}$$

where $b_m > 0$ is a design constant and, here and throughout, $[u]$ will stand for the greatest integer less than u. Next, take an additional sample $X_{m+1}, ..., X_N$ of size $N - m$ accordingly and calculate T_N based on all the samples. Then, the fixed-width confidence interval for θ is finally obtained as $[T_N - d,\ T_N + d]$, keeping in mind that the lower or upper confidence limit will be modified suitably if the domain space depends on θ. Let us verify satisfaction of the probability requirement. Noting the stopping rule (1) and assumptions (a)-(b), we have

$$\begin{aligned} P(|T_N - \theta| \le d) &= P(N^\gamma |T_N - \theta|/g(\xi) \le N^\gamma d/g(\xi)) \\ &\ge P(m^\gamma |T_m - \theta|/g(\xi) \le b_m g(U_m)/g(\xi)). \end{aligned}$$

So, the probability requirement is satisfied by designing b_m as a constant such that

$$P(m^\gamma |T_m - \theta|/g(U_m) \le b_m) = 1 - \alpha.$$

By the way, when b_m is designed as the positive square root of

$$b_m^2 = E\{(m^\gamma |T_m - \theta|/g(U_m))^2\},$$

we have the following about a risk of T_N:

$$\begin{aligned} E(|T_N - \theta|^2) &= E\{(N^\gamma |T_N - \theta|/g(\xi))^2 (g(\xi)/N^\gamma)^2\} \\ &\le (d/b_m)^2 E\{(m^\gamma |T_m - \theta|/g(\xi))^2 (g(\xi)/g(U_m))^2\} = d^2. \end{aligned}$$

It should be noted that the risk is bounded above by the preassigned number $d\ (> 0)$.

For example, suppose the X's are from $N(\mu, \sigma^2)$. Choose $\theta = \mu,\ \xi = \sigma$ and for $n \ge 4$ let $T_n = \overline{X}_n\ (= n^{-1} \sum_{i=1}^n X_i)$, $U_n = S_n\ (= (n-1)^{-1/2} (\sum_{i=1}^n (X_i - \overline{X}_n)^2)^{1/2})$. In this case, set $\gamma = 1/2$ and $g(x) = x\ (> 0)$. Then, assumptions (a)-(b) are satisfied. Finally, Stein's two-stage procedure yields the fixed-width confidence interval $[\overline{X}_N - d,\ \overline{X}_N + d]$ with confidence $1 - \alpha$ when b_m is designed as the upper $\alpha/2$ point of Student's t-distribution with $m - 1$ degrees of freedom (d.f.), while the estimator $\overline{X}_N$ has a risk bounded above by d^2 when b_m is designed as the standard deviation of the same t-distribution. For another example, we may consider a case in which the domain space depends on the parameter to be estimated. Suppose that the X's are from a negative exponential distribution $e(\mu, \sigma)$ with the p.d.f.

$$f(x; \mu, \sigma) = \sigma^{-1} \exp\left(-\frac{x - \mu}{\sigma}\right) I(x > \mu),$$

where $I(\cdot)$ stands for the indicator function of $(\cdot)$. Choose $\theta = \mu,\ \xi = \sigma$, and for $n \ge 4$ let $T_n = \min(X_1, ..., X_n)$, $U_n = (n-1)^{-1} \sum_{i=1}^n (X_i - T_n)$. In this case, set $\gamma = 1$ and $g(x) = x\ (> 0)$. Then, assumptions (a)-(b) are satisfied. Exchange d in the stopping rule (1) with $2d$ and define the fixed-width confidence interval for μ by $[T_N - 2d,\ T_N]$. That interval has confidence $1 - \alpha$ when b_m is designed as the upper α point of the F-distribution with $(2,\ 2(m-1))$ d.f.s, while the risk of T_N is bounded above by $4d^2$ when b_m is designed as the positive square root of the second order moment for the same F-distribution.

Several estimation problems for parameters such as the location parameter of a negative exponential distribution, the scale parameter of a Pareto distribution, and the mean of an inverse Gaussian distribution have the same nature as for the mean of a normal distribution in the sense of the essence of Stein's two-stage procedure. Therefore, it is possible to apply the methodologies in this article to these problems. See Ghurye [G58] and Mukhopadhyay [M82] for the details. In addition, the estimation problem for regression parameters of a linear regression model, and the ranking and selection problem of populations can be considered in a similar context.

Even if the essence of the two–stage procedure in the one-dimensional case of a population distribution remains in the multi-dimensional case, still, handling the statistic that appears in the multi-dimensional case becomes complicated. For problems of processing several populations simultaneously, it becomes even more complicated. The efficiency of inference really depends on how you handle properties of the underlying distribution in methodologies for an individual problem. In all problems, the two–stage procedure aims at guaranteeing a requirement about a risk in authenticity (instead of approximation). With that in mind, we pursue an optimal way to reduce the sample size required in inference. In Sections 3-4, a multivariate normal distribution is taken as a population distribution, so that we are not bothered by specific details peculiar to distribution. We shall solve an individual inference problem by using a two–stage procedure. There are several cases when assumptions (a)-(b) are not satisfied. In Section 5, about inference problems for such cases, we shall discuss robustness of a two–stage procedure, along with several other techniques to handle such cases.

3. A multivariate normal population

Let $\{\boldsymbol{X}_i;\ i \geq 1\}$ be a sequence of i.i.d. random vectors with p-variate normal distribution $N_p(\boldsymbol{\mu}, \boldsymbol{\Sigma})$, where $\boldsymbol{\Sigma} > \boldsymbol{0}$. Let us consider the following two inference problems about the mean vector $\boldsymbol{\mu}$.

(P1) For a prespecified constant $\varepsilon > 0$, find an estimator $\boldsymbol{\delta}_n = \boldsymbol{\delta}_n(X_1, ..., X_n)$ for $\boldsymbol{\mu}$ such that $E(||\boldsymbol{\delta}_n - \boldsymbol{\mu}||^2) \leq \varepsilon$ for any $(\boldsymbol{\mu}, \boldsymbol{\Sigma})$. Here and throughout, the norm is the Euclidian norm and $||\boldsymbol{X}||^2 = \boldsymbol{X}'\boldsymbol{X}$, where $\boldsymbol{X}'$ stands for the transpose of a vector $\boldsymbol{X}$.

(P2) For a value $d > 0$ of the radius and a confidence coefficient $1 - \alpha > 0$, both given beforehand, determine the sample size n such that the confidence region $CR(\boldsymbol{\delta}_n) = \{\boldsymbol{\mu}|\ ||\boldsymbol{\mu} - \boldsymbol{\delta}_n|| \leq d\}$ for $\boldsymbol{\mu}$ satisfies $P(\boldsymbol{\mu} \in CR(\boldsymbol{\delta}_n)) \geq 1 - \alpha$ for any $(\boldsymbol{\mu}, \boldsymbol{\Sigma})$.

Having recorded $\boldsymbol{X}_1, ..., \boldsymbol{X}_n$ of size n, it is natural to consider the sample mean $\overline{\boldsymbol{X}}_n$ as an estimator of $\boldsymbol{\mu}$. When $\boldsymbol{\Sigma}$ is known, we have

$$E(||\overline{\boldsymbol{X}}_n - \boldsymbol{\mu}||^2) = tr(\boldsymbol{\Sigma})/n,$$

so that the sample size n needed to solve (P1) is determined as the smallest integer such that $n \geq tr(\boldsymbol{\Sigma})/\varepsilon$. As for (P2), let a be the upper α point of the chi-squared distribution function, $F_p(\cdot)$, with p d.f., i.e., $F_p(a) = 1 - \alpha$. Consider the confidence region $CR(\overline{\boldsymbol{X}}_n)$. When $\boldsymbol{\Sigma}$ is known, the sample size n needed to solve (P2) is determined as the smallest integer such that $n \geq a\lambda/d^2$, where λ denotes the maximum latent root of $\boldsymbol{\Sigma}$. However, if $\boldsymbol{\Sigma}$ is unknown, the sample size n should be a function of a sample through an estimate of $\boldsymbol{\Sigma}$. To handle this process, the two–stage procedure comes into consideration. As for (P1), it is solved as follows.

[Two–stage procedure (P1)]

(T1) First, take a pilot sample $\boldsymbol{X}_1, ..., \boldsymbol{X}_m$ of size m (≥ 4) and calculate $\boldsymbol{S}_m = (m-1)^{-1}\sum_{i=1}^{m}(\boldsymbol{X}_i - \overline{\boldsymbol{X}}_m)(\boldsymbol{X}_i - \overline{\boldsymbol{X}}_m)'$ as an estimate of $\boldsymbol{\Sigma}$. Define the stopping rule

$$(2) \qquad N = \max\{m,\ [c_m tr(\boldsymbol{S}_m)/\varepsilon] + 1\}, \quad \text{where } c_m = (m-1)/(m-3).$$

(T2) Next, take an additional sample $\boldsymbol{X}_{m+1}, ..., \boldsymbol{X}_N$ of size $N-m$. By combining the initial sample and the additional sample, estimate $\boldsymbol{\mu}$ by $\overline{\boldsymbol{X}}_N = N^{-1}\sum_{i=1}^N \boldsymbol{X}_i$.

When $p = 1$, Birnbaum and Healy [BH60] showed that the above two–stage procedure solves problem (P1), and it was extended to the case when $p \geq 2$ by Kubokawa [K90]. (Refer to Rao [Ra73, pp. 486–487] as well.) As for a choice of the size m, Cohen and Sackrowitz [CS84ab] studied the Bayes decision rule with respect to suitable prior distributions of the parameters.

When we assume that we have a known real number $s_\star$ such that $tr(\boldsymbol{\Sigma}) > s_\star (> 0)$ for any $\boldsymbol{\Sigma}$ $(> \boldsymbol{0})$, Aoshima [A00] showed that procedure (P1) is second–order efficient under the condition that $m = m(\varepsilon)$ and $\lim_{\varepsilon \to 0} m\varepsilon = s_\star$, by developing the techniques given by Mukhopadhyay and Duggan [MD97, 99]. This means that the average sample size and the associated risk can be expanded up to the second order when $\varepsilon \to 0$. It would be interesting if a reduction of the sample size, required in (P1), is considered with the help of improvements on estimators together with the stopping rule by incorporating prior information about nuisance parameters.

For the specfic case $\boldsymbol{\Sigma} = \sigma^2 \boldsymbol{I}_p$, Ghosh and Sen [GS83] showed that the James-Stein type estimator

$$\boldsymbol{\delta}_N = \overline{\boldsymbol{X}}_N - \frac{(p-2)\hat{\sigma}_N^2}{N||\overline{\boldsymbol{X}}_N||^2}\overline{\boldsymbol{X}}_N, \quad \hat{\sigma}_N^2 = \sum_{i=1}^N ||\boldsymbol{X}_i - \overline{\boldsymbol{X}}_N||^2/\{p(N-1)+2\}$$

dominates $\overline{\boldsymbol{X}}_N$ when $p \geq 3$. As for a purely sequential procedure, Ghosh et al. [GNS87] showed a similar risk dominance result. Natarajan and Strawderman [NS85] and Kubokawa and Saleh [KS94] studied the improvements of the stopping rule together with the estimator such that in the shrinkage procedures the sample size is less than or equal to N and the shrinkage estimator is asymptotically better than $\overline{\boldsymbol{X}}_N$ when $\boldsymbol{\Sigma} = \sigma^2 \boldsymbol{I}_p$ with $p \geq 3$. As for the case of arbitrary $\boldsymbol{\Sigma}$, the James-Stein type estimator has not been developed.

When certain prior distributions are assumed for the parameters, the Bayes sequential estimation is pursued to minimize the Bayes risk over all stopping rules and over all estimators. Arrow et al. [ABG49] showed that in the Bayes sequential estimation problem, the optimal estimator for squared loss is given by the Bayes solution for any stopping rule. It can be shown that an optimal stopping rule exists; however it is given by the method of backward induction and it is often inaccessible. To overcome this difficulty, Bickel and Yahav [BY68] devised the APO (asymptotically pointwise optimal) rule, which derives an explicit stopping rule whose Bayes risk typically is close to the Bayes risk of the optimal rule. When a conjugate distribution is taken as a prior distribution, the APO rule has a risk of the same extent as the Bayes stopping rule asymptotically as the observation cost approaches zero. When $p = 1$, Woodroofe [W81] showed that certain stopping rules are asymptotically non-deficient as the observation cost approaches zero: That verification was extended to the multivariate case by Nagao [N97ab]. When the prior distribution is not completely known but auxiliary data are available for estimating unknown parameters of the prior distribution, Martinsek [M87] considered a general empirical Bayes approach which approximates the optimal rule. On the other hand, to overcome the difficulty, Alvo [Alv77, 78], Akahira and Koike [AkK96], Koike [Ko99] and others considered a heuristic approach giving stopping rules for which the excess risk incurred over the optimum Bayes risk is bounded and possibly evaluated explicitly from the prior distribution. Further evolutions, such as improvements of the inequality so that a certain distribution family achieving the bound exists, could be anticipated in this research.

As for (P2), it is solved as follows.

[Two–stage procedure (P2)]

(T1) First, take a pilot sample $\boldsymbol{X}_1, ..., \boldsymbol{X}_m$ of size m ($> p$) and calculate the maximum latent root ℓ_m of $\boldsymbol{S}_m$. Define the stopping rule

$$N = \max\left\{m, \ \left[a_m \ell_m / d^2\right] + 1\right\}, \tag{3}$$

where $a_m = p(m-1)F_{p,m-p}(\alpha)/(m-p)$ with $F_{r,s}(\alpha)$ the upper α point of the F distribution having d.f.s (r, s).

(T2) Next, take an additional sample $\boldsymbol{X}_{m+1}, ..., \boldsymbol{X}_N$ of size $N - m$. By combining the initial sample and the additional sample, define the confidence region $CR(\overline{\boldsymbol{X}}_N)$ with $\overline{\boldsymbol{X}}_N = N^{-1}\sum_{i=1}^N \boldsymbol{X}_i$.

Healy [He56] showed that the above two–stage procedure solves problem (P2). Aoshima [A00] suggested that the stopping rule (3) replacing a_m by the smaller constant $a_m^\star = pF_{p,m-1}(\alpha)$ still enables the procedure to solve problem (P2). When p is large and m is small, the reduction of N by this correction is remarkable. Under the assumptions that λ is simple and there exists a known and positive $\lambda_\star$ such that $\lambda > \lambda_\star$ for any $\boldsymbol{\Sigma}$ ($> \boldsymbol{0}$), Mukhopadhyay [M99] and [A00] considered a sequence of $m = m(d)$ such that $\lim_{d\to 0} md^2 = a\lambda_\star$ and studied the second–order efficiency of procedure (P2) as $d \to 0$. Recall that a denotes the upper α point of the chi–squared distribution with p d.f. As for a purely sequential procedure for problem (P2), see Srivastava [Sriv67], and refer to Woodroofe [W77] for its second–order efficiency when $p = 1$: For a multivariate case, its second–order efficiency has not been resolved fully, but Dmitrienko and Govindarajulu [DB00] proved risk boundedness.

It would be interesting to consider a James-Stein type estimator improving $CR(\overline{\boldsymbol{X}}_N)$ at least when $\boldsymbol{\Sigma} = \sigma^2 \boldsymbol{I}_p$. For this challenging problem, Hwang and Casella [HC82, 84], Shinozaki [Sh89] and Takada [T98b] should be consulted: They considered this issue in a single–stage sampling scheme. In addition, about an empirical Bayes confidence region, it has not been elucidated enough theoretically even in a single–stage sampling scheme.

Another multivariate version of Stein's [S45] two–stage procedure is available for (P2).

[Two–stage procedure (P2)′]

(T1) First, take a pilot sample $\boldsymbol{X}_1, ..., \boldsymbol{X}_m$ of size m ($> p$) and calculate $tr(\boldsymbol{S}_m)$. Define the stopping rule

$$\tilde{N} = \max\left\{m + p^2, \ \left[\tilde{a}_m tr(\boldsymbol{S}_m)/d^2\right] + 1\right\}, \tag{4}$$

where $\tilde{a}_m > 0$ is given as $u = \tilde{a}_m$ such that $\int_0^\infty v^{-1} f_p(v^{-1}u) g_{p,m}(v) dv = 1 - \alpha$ with $f_p(\cdot)$ p.d.f. of the chi-squared distribution with p d.f. and $g_{p,m}(\cdot)$ p.d.f. of $V = tr(\boldsymbol{\Sigma}\boldsymbol{S}_m^{-1})$.

(T2) Next, take an additional sample $\boldsymbol{X}_{m+1}, ..., \boldsymbol{X}_{\tilde{N}}$ of size $\tilde{N} - m$. By combining the initial sample and the additional sample, calculate a generalized sample mean vector $\widetilde{\boldsymbol{X}}_{\tilde{N}} = (tr(\boldsymbol{T}_1\boldsymbol{Y}'), ..., tr(\boldsymbol{T}_p\boldsymbol{Y}'))'$, where $\boldsymbol{Y} = [\boldsymbol{X}_1, ..., \boldsymbol{X}_m, \boldsymbol{X}_{m+1}, ..., \boldsymbol{X}_{\tilde{N}}]$ and $\boldsymbol{T}_i = [\boldsymbol{t}_{i1}, ..., \boldsymbol{t}_{im}, \boldsymbol{t}_{im+1}, ..., \boldsymbol{t}_{i\tilde{N}}]$, $1 \le i \le p$, are $p \times \tilde{N}$ random matrices satisfying the following three conditions:

(i) $\boldsymbol{t}_{i1} = ... = \boldsymbol{t}_{im}$;

(ii) $\boldsymbol{T}_i\boldsymbol{1} = \boldsymbol{e}_i$, where $\boldsymbol{1} : \tilde{N} \times 1 = (1, ..., 1)'$ and $\boldsymbol{e}_i : p \times 1 = (0, ..., 0, 1, 0, ..., 0)'$;
(iii) $\boldsymbol{T}\boldsymbol{T}' = (d^2/\tilde{a}_m)\boldsymbol{I}_p \otimes \boldsymbol{S}_m^{-1}$, where $\boldsymbol{T} : p^2 \times \tilde{N} = [\boldsymbol{T}_1', ..., \boldsymbol{T}_p']'$.

Then, define the confidence region $CR(\widetilde{\boldsymbol{X}}_{\tilde{N}})$ with $\widetilde{\boldsymbol{X}}_{\tilde{N}}$.

The conditional distribution of $\widetilde{\boldsymbol{X}}_{\tilde{N}}$ given $\boldsymbol{S}_m$ is $N_p(\boldsymbol{\mu}, (Vd^2/\tilde{a}_m)\boldsymbol{I}_p)$. When $p = 1$, Stein [S45] showed that $\sqrt{\tilde{a}_m/(pd^2)}(\widetilde{\boldsymbol{X}}_{\tilde{N}} - \boldsymbol{\mu})$ has Student's t-distribution with $\nu = m - 1$ d.f. When $p = 2$, Chatterjee [C59] gave its distribution. When $p \geq 3$, the distribution of $\widetilde{\boldsymbol{X}}_{\tilde{N}}$ is complicated to handle exactly. In this case, a large sample approximation could be considered. Hyakutake and Siotani [HS87] obtained an asymptotic expansion for the distribution of $\sqrt{\tilde{a}_m/(pd^2)}(\widetilde{\boldsymbol{X}}_{\tilde{N}} - \boldsymbol{\mu})$ up to the order $O(\nu^{-2})$ by using the differential operator method. Then, its limiting distribution is $N_p(\boldsymbol{0}, \boldsymbol{I}_p)$.

Chatterjee [C60] showed that the above two–stage procedure also solves problem (P2) with coverage probability exactly equal to $1 - \alpha$. The stopping rule (4) is the least integer meeting the necessary and sufficient condition to choose $\boldsymbol{T}$ satisfying three conditions (i)–(iii) simultaneously. (Several methods of generating $\boldsymbol{T}$ were given by Hyakutake [H86] and Dudewicz and Taneja [DT87].) Chatterjee [C60] showed that procedure (P2)$'$ is less efficient than (P2) in terms of both the sample size and the coverage probability. Furthermore, for any $(\boldsymbol{\mu}, \boldsymbol{\Sigma})$, we have $CR(\overline{\boldsymbol{X}}_N) \subset CR(\widetilde{\boldsymbol{X}}_{\tilde{N}})$ with probability 1 when placed at the same center. (Aoshima [A94a] gave assessments about both mean and variance of the sample size in procedure (P2) and compared them numerically with those given by Hyakutake and Siotani [HS89] in (P2)$'$.) However, on the other hand, procedure (P2)$'$ has the interesting property that $U = \tilde{a}_m(pd^2)^{-1}||\widetilde{\boldsymbol{X}}_{\tilde{N}}||^2$ has distribution depending only on $||\boldsymbol{\mu}||^2$ but completely free from nuisance parameters $\boldsymbol{\Sigma}$. Chatterjee [C59] successfully showed that procedure (P2)$'$ yields a test for $H_0 : \boldsymbol{\mu} = \boldsymbol{0}$ with power function completely free from $\boldsymbol{\Sigma}$. (As for the distribution of U, an asymptotic expansion was studied by Mukaihata and Fujikoshi [MF93] together with its error bound.) Albers [Alb92] investigated to what extent this nice feature of having a power independent of certain aspects of the unknown underlying distribution can be generalized to the case of rank tests.

It is not known whether the test given by procedure (P2) is uniformly more powerful than the test given by (P2)$'$. Chatterjee [C60] constructed a test which is uniformly more powerful and slightly more economical than (P2)$'$, by replacing $m + p^2$ with $m + p^2 - 1$ in the stopping rule (4) and by replacing $(d^2/\tilde{a}_m)$ with $tr(\boldsymbol{S}_m)/\tilde{N}$ in condition (iii) about $\boldsymbol{T}$. (Note that for $\tilde{N}$ given by those replacements such a $\boldsymbol{T}$ can always be found.) It is evident that procedure (P2)$'$ with the above modification can also yield a confidence region slightly better than $CR(\widetilde{\boldsymbol{X}}_{\tilde{N}})$. However, as suggested in [C60], this slight improvement requires the sacrifice that the associated power already depends on $\boldsymbol{\Sigma}$. See also Chattejee [C91]. As for a purely sequential procedure, refer to Liu [Li97c], who studied a test problem when $p = 1$.

When some structure of $\boldsymbol{\Sigma}$ is supposed by prior information, several improvements of the stopping rule can be considered. For instance, let us consider the case that $\boldsymbol{\Sigma}$ has an intraclass correlation model such as $\boldsymbol{\Sigma} = \sigma^2\{(1-\rho)\boldsymbol{I}_p + \rho\boldsymbol{1}\boldsymbol{1}'\}$, where $\boldsymbol{1} = (1, ..., 1)'$. Then, as for the maximum latent root of $\boldsymbol{\Sigma}$, we have $\lambda = \max(\tau_1, \tau_2)$, where $\tau_1 = \sigma^2\{1 + (p-1)\rho\}$ is simple while $\tau_2 = \sigma^2(1-\rho)$ is $p - 1$ multiple. Note that $\hat{\tau}_{1m} = p^{-1}\boldsymbol{1}'\boldsymbol{S}_m\boldsymbol{1}$ and $\hat{\tau}_{2m} = (p-1)^{-1}(tr(\boldsymbol{S}_m) - \hat{\tau}_{1m})$ are unbiased estimates of τ_1 and τ_2 respectively. A natual estimation of λ could be $\max(\hat{\tau}_1, \hat{\tau}_2)$. Hyakutake et al. [HTA95] considered an improvement of procedure (P2) by modifying the stopping rule (3) as

$$N = \max\left\{m,\ \left[a_m \max(\hat{\tau}_{1m}, \hat{\tau}_{2m})/d^2\right] + 1\right\}. \tag{5}$$

Problem (P2) can be solved by choosing a_m as a constant such that

$$P\{U_1 + (p-1)U_2 \le a_m\} = 1 - \alpha,$$

where U_1 and U_2 are independent F-random variables with $(1, m-1)$ d.f.s and $(p-1, (p-1)(m-1))$ d.f.s, respectively. As expected, for that structure of $\boldsymbol{\Sigma}$, procedure (P2) with (3) is improved by the modification with (5) in terms of the sample size. Takada and Hyakutake [TH97] considered a similar modification under the generalization that $\boldsymbol{\Sigma} = \sigma_1 \boldsymbol{A}_1 + ... + \sigma_q \boldsymbol{A}_q$, where $\boldsymbol{A}_i$ is a known and symmetric matrix with $rank(\boldsymbol{A}_i) = r_i$ for each i and $\sum_{i=1}^q \boldsymbol{A}_i = \boldsymbol{I}_p$, $\sum_{i=1}^q r_i = p$, while the σ_i's are unknown and positive scalars. Under that general structure of $\boldsymbol{\Sigma}$, Aoshima and Mukhopadhyay [AM99] gave asymptotic assessments up to the second order when $d \to 0$ about the average sample size and the coverage probability for the modified procedure (P2). For a purely sequential procedure, see Nagao [N96].

4. Several multivariate normal populations

Let us consider estimating a linear function of mean vectors coming from several multivariate normal populations. For each i $(1 \le i \le k)$, let $\{\boldsymbol{X}_{ij};\ j \ge 1\}$ be a sequence of i.i.d. random vectors with p-variate normal distribution $N_p(\boldsymbol{\mu}_i, \boldsymbol{\Sigma}_i)$, where $\boldsymbol{\Sigma}_i > \boldsymbol{0}$. For $(b_1, ..., b_k)$ given in advance for the aim of inference, we define $\boldsymbol{\xi} = \sum_{i=1}^k b_i \boldsymbol{\mu}_i$. Let us consider the following two inference problems about $\boldsymbol{\xi}$.

(P3) For a prespecified constant $\varepsilon > 0$, find $\boldsymbol{n} = (n_1, ..., n_k)$, the sample sizes from each population, and an estimator $\boldsymbol{\delta}_{\mathbf{n}}$ of $\boldsymbol{\xi}$ such that $E(||\boldsymbol{\delta}_{\mathbf{n}} - \boldsymbol{\xi}||^2) \le \varepsilon$ for any $(\boldsymbol{\mu}_i, \boldsymbol{\Sigma}_i)$, $1 \le i \le k$.

(P4) For a value $d > 0$ of the radius and a confidence coefficient $1 - \alpha > 0$, both given beforehand, determine $\boldsymbol{n} = (n_1, ..., n_k)$ such that the confidence region $CR(\boldsymbol{\delta}_{\mathbf{n}}) = \{\boldsymbol{\xi} |\ ||\boldsymbol{\xi} - \boldsymbol{\delta}_{\mathbf{n}}|| \le d\}$ for $\boldsymbol{\xi}$ satisfies $P(\boldsymbol{\xi} \in CR(\boldsymbol{\delta}_{\mathbf{n}})) \ge 1 - \alpha$ for any $(\boldsymbol{\mu}_i, \boldsymbol{\Sigma}_i)$, $1 \le i \le k$.

Having recorded $\boldsymbol{X}_{i1}, ..., \boldsymbol{X}_{in_i}$ of size n_i for each i $(1 \le i \le k)$, we define the linear function of sample means by $\boldsymbol{T}_{\mathbf{n}} = \sum_{i=1}^k b_i \overline{\boldsymbol{X}}_{in_i}$. For (P3), we have

$$E(||\boldsymbol{T}_{\mathbf{n}} - \boldsymbol{\xi}||^2) = \sum_{i=1}^k b_i^2 tr(\boldsymbol{\Sigma}_i)/n_i.$$

When $\boldsymbol{\Sigma}_i$ $(1 \le i \le k)$ is known, the sample sizes $\boldsymbol{n}$ to solve (P3) would be determined as the smallest integer such that

$$n_i \ge |b_i|\sqrt{tr(\boldsymbol{\Sigma}_i)} \sum_{i=1}^k |b_i|\sqrt{tr(\boldsymbol{\Sigma}_i)}/\varepsilon$$

for each i if one considers minimizing the total sample size $\sum_{i=1}^k n_i$. For (P4), the sample sizes $\boldsymbol{n}$ are determined as the smallest integer such that

$$n_i \ge a|b_i|\sqrt{\lambda_i} \sum_{i=1}^k |b_i|\sqrt{\lambda_i}/d^2$$

for each i to construct $CR(\boldsymbol{T}_{\mathbf{n}})$. Here, a is the upper α point of the chi-squared distribution with p d.f., and λ_i denotes the maximum latent root of $\boldsymbol{\Sigma}_i$. However, if $\boldsymbol{\Sigma}_i$ $(1 \le i \le k)$ is unknown, the sample sizes $\boldsymbol{n}$ should be determined through estimates of $\boldsymbol{\Sigma}_i$, $1 \le i \le k$. As for (P3), it is solved as follows.

[Two–stage procedure (P3)]

(T1) First, take a pilot sample $\boldsymbol{X}_{i1}, ..., \boldsymbol{X}_{im}$ of size m (≥ 4) and calculate $\boldsymbol{S}_{im} = (m-1)^{-1}\sum_{j=1}^{m}(\boldsymbol{X}_{ij} - \overline{\boldsymbol{X}}_{im})(\boldsymbol{X}_{ij} - \overline{\boldsymbol{X}}_{im})'$ as an estimate of $\boldsymbol{\Sigma}_i$ for each i. Define the stopping rule of each population by

$$N_i = \max\left\{m,\ \left[c_m|b_i|\sqrt{tr(\boldsymbol{S}_{im})}\sum_{i=1}^{k}|b_i|\sqrt{tr(\boldsymbol{S}_{im})}/\varepsilon\right]+1\right\}, \tag{6}$$

where $c_m = (m-1)/(m-3)$, which is the same one as in (2).

(T2) Next, take an additional sample $\boldsymbol{X}_{im+1}, ..., \boldsymbol{X}_{iN_i}$ of size $N_i - m$ for each i. Let $\boldsymbol{N} = (N_1, ..., N_k)$. By combining the initial sample and the additional sample, calculate $\overline{\boldsymbol{X}}_{iN_i} = N_i^{-1}\sum_{j=1}^{N_i}\boldsymbol{X}_{ij}$ for each i. Then, estimate $\boldsymbol{\xi}$ by $\boldsymbol{T}_{\mathbf{N}} = \sum_{i=1}^{k} b_i\overline{\boldsymbol{X}}_{iN_i}$.

Aoshima and Takada [AT02] showed that the above two–stage procedure solves problem (P3) and improves the predecessor research given in Ghosh et al. [GMS97, Chap. 6] in terms of the sample size.

As for (P4), it is solved as follows.

[Two–stage procedure (P4)]

(T1) First, take a pilot sample $\boldsymbol{X}_{i1}, ..., \boldsymbol{X}_{im}$ of size m $(> p)$ and calculate the maximum latent root ℓ_{im} of $\boldsymbol{S}_{im}$ for each i. Define the stopping rule of each population by

$$N_i = \max\left\{m,\ \left[u_m|b_i|\sqrt{\ell_{im}}\sum_{i=1}^{k}|b_i|\sqrt{\ell_{im}}/d^2\right]+1\right\}, \tag{7}$$

where $u_m > 0$ is determined by solving the equations

$$kF_{p,m-1}\left(\frac{u_m}{p}\right) - (k-1)F_{p,m}\left(\frac{u_m m}{p(m-1)}\right) = 1-\alpha$$

when $p = 1, 2$, and

$$k\int_0^\infty F_p\left(\frac{u_m x}{m-1}\right)(1 - F_{m-1}(x))^{k-1}dF_{m-1}(x) = 1-\alpha$$

when $p \ge 3$. Here, $F_{r,s}(\cdot)$ and $F_r(\cdot)$ denote the c.d.f.s of the F-distribution with (r, s) d.f.s. and chi-squared distribution with r d.f., respectively.

(T2) Next, take an additional sample $\boldsymbol{X}_{im+1}, ..., \boldsymbol{X}_{iN_i}$ of size $N_i - m$ for each i. Let $\boldsymbol{N} = (N_1, ..., N_k)$. By combining the initial sample and the additional sample, calculate $\overline{\boldsymbol{X}}_{iN_i} = N_i^{-1}\sum_{j=1}^{N_i}\boldsymbol{X}_{ij}$ for each i. Then, define the confidence region $CR(\boldsymbol{T}_{\mathbf{N}})$ with $\boldsymbol{T}_{\mathbf{N}} = \sum_{i=1}^{k} b_i\overline{\boldsymbol{X}}_{iN_i}$.

Aoshima et al. [ATS02] showed that the above two–stage procedure solves problem (P4). Problem (P4) had been studied by Chapman [Ch50], Ghosh [Gh75], Mukhopadhyay and Liberman [ML89] and many other researchers, but an optimal solution had not been obtained even asymptotically. As for specific cases of (P4), when $p = 1$ and $k = 2$, Banerjee [B67] gave an asymptotic optimal solution and Schwabe [Sc93] gave an improvement of [B67] in terms of the sample size. Takada and Aoshima [TA96] developed the Banerjee-Schwabe result for a multivariate case but with $\boldsymbol{\Sigma}_i = \sigma_i^2 \boldsymbol{H}_i$ ($\boldsymbol{H}_i$ is a known and positive definite matrix), $i = 1, 2$, and this was followed by Takada and Aoshima [TA97] for the case when $k \geq 3$. Following these papers, [ATS02] gave an asymptotic optimal solution to problem (P4) in which $\boldsymbol{\Sigma}_i$ $(1 \leq i \leq k)$ is completely unknown. For second–order asymptotic properties in this context, see Aoshima and Mukhopadhyay [AM02]. Assuming for each i that $\boldsymbol{\Sigma}_i = \sigma_{i1}\boldsymbol{A}_1 + ... + \sigma_{iq}\boldsymbol{A}_q$ with known and symmetric $\boldsymbol{A}_j$ $(1 \leq j \leq q)$ such that $rank(\boldsymbol{A}_j) = r_j$, $\sum_{j=1}^q \boldsymbol{A}_j = \boldsymbol{I}_p$, $\sum_{j=1}^q r_j = p$ and with unknown and positive scalars σ_{ij} $(1 \leq j \leq q)$, Aoshima [A98] and Hyakutake [H98] studied some modifications of the stopping rule (7).

When $k = 2$ and $(b_1, b_2) = (1, -1)$, problem (P4) reduces to the Behrens-Fisher problem. Then, for testing $H_0 : \boldsymbol{\mu}_1 = \boldsymbol{\mu}_2$, procedure (P2)$'$ given in Section 3 could be considered naturally. Dudewicz and Bishop [DB79] proposed an application of procedure (P2)$'$ to the Behrens-Fisher problem, noting that the variance-covariance matrices of the generalized sample mean vectors given in procedure (P2)$'$ are common and known for two populations if one conducts procedure (P2)$'$ for each population independently but with a common size of m $(> p)$ for their pilot samples. Sample sizes for that testing are determined so that a test rule with significant level $\alpha > 0$ against $H_1 : \boldsymbol{\mu}_1 \neq \boldsymbol{\mu}_2$ guarantees power β when $||\boldsymbol{\mu}_1 - \boldsymbol{\mu}_2|| = \delta_0$ for given $\delta_0 > 0$ and $\beta > 0$. For constructing simultaneous confidence intervals, see Hyakutake and Siotani [HS87], Siotani [Sio87], Siotani et al. [SHF85; Secs. 5.6.3, 6.4.3] and Aoshima [A94b], and the references given in them.

When $p \geq 2$, we consider multiple comparison experiments for correlated components $(\xi_1, ..., \xi_p)$ of $\boldsymbol{\xi} = \sum_{i=1}^k b_i \boldsymbol{\mu}_i$. When we suppose that there are several remedies to be compared with each other, say $k = 2$, and those effects are observed at p points of a time series, the user would be typically interested in the direction and the magnitude of differences—which points are more significant differences, and by how much—with respect to p correlated time components $(\xi_1, ..., \xi_p)$ of $\boldsymbol{\mu}_1 - \boldsymbol{\mu}_2$ $(b_1 = 1, b_2 = -1)$. For multiple comparison methods, Tukey's (1953) method of all pairwise multiple comparisons (MCA), Hsu's (1984) method of multiple comparisons with the best (MCB), and Dunnett's (1955) method of multiple comparisons with a control (MCC) are well known and referred to by many authors. (See Hochberg and Tamhane [HT87], Hirotsu [Hi92], Hsu [Hs96], and Nagata and Yoshida [NY97] as related textbooks.) Having recorded $\boldsymbol{X}_{i1}, ..., \boldsymbol{X}_{in_i}$ from each i $(1 \leq i \leq k)$, we have $\boldsymbol{T}_{\mathbf{n}} = \sum_{i=1}^k b_i \overline{\boldsymbol{X}}_{in_i}$ with $\boldsymbol{n} = (n_1, ..., n_k)$. Then, the components $(\xi_1, ..., \xi_p)$ are estimated by $\boldsymbol{T}_{\mathbf{n}} = (T_{1\mathbf{n}}, ..., T_{p\mathbf{n}})$. Tukey's MCA method, Hsu's MCB method and Dunnett's MCC method give the following simultaneous confidence intervals when d is specified (suitably narrow).

(MCA) For the $p(p-1)/2$ differences of component effects,

$$SCI(\boldsymbol{T}_{\mathbf{n}}) = \{\boldsymbol{\xi} |\ \xi_r - \xi_s \in [T_{r\mathbf{n}} - T_{s\mathbf{n}} - d,\ T_{r\mathbf{n}} - T_{s\mathbf{n}} + d],\ 1 \leq r < s \leq p\}.$$

(MCB) For comparing each component with the best of the other components when a larger component effect is supposed to be better,

$$\begin{aligned} &SCI(\boldsymbol{T}_{\mathbf{n}}) \\ &\quad = \{\boldsymbol{\xi} |\ \xi_r - \max_{s \neq r} \xi_s \in [-(T_{r\mathbf{n}} - \max_{s \neq r} T_{s\mathbf{n}} - d)^-,\ +(T_{r\mathbf{n}} - \max_{s \neq r} T_{s\mathbf{n}} + d)^+], \\ &\qquad\qquad r = 1, ..., p\}, \end{aligned}$$

where $+x^+ = \max\{0, x\}$ and $-x^- = \min\{0, x\}$.

(MCC) For comparing each component with a control component,

$$SCI(\boldsymbol{T}_{\mathbf{n}}) = \{\boldsymbol{\xi} |\ \xi_r - \xi_p \in [T_{r\mathbf{n}} - T_{p\mathbf{n}} - d,\ T_{r\mathbf{n}} - T_{p\mathbf{n}} + d],\ r = 1, ..., p-1\},$$

where the component p is the control.

Then, we consider the following problem for each of (MCA), (MCB) and (MCC).

(P5) For a value $d > 0$ and a confidence coefficient $1 - \alpha > 0$ both given beforehand, determine $\boldsymbol{n} = (n_1, ..., n_k)$ such that the simultaneous confidence intervals $SCI(\boldsymbol{T}_{\mathbf{n}})$ satisfy $P(\boldsymbol{\xi} \in SCI(\boldsymbol{T}_{\mathbf{n}})) \geq 1 - \alpha$ for any $(\boldsymbol{\mu}_i, \boldsymbol{\Sigma}_i)$, $1 \leq i \leq k$.

The coverage probability of simultaneous confidence intervals depends on the variance-covariance structure only through the variance of the pairwise differences of the component mean estimates. We assume that $\boldsymbol{\Sigma}_i = (\sigma_{(i)rs})$ has a spherical structure such as $\sigma_{(i)rr} + \sigma_{(i)ss} - 2\sigma_{(i)rs} = 2\tau_i^2$ $(1 \leq r < s \leq p)$ depending only on the suffix i $(1 \leq i \leq k)$. (See Hirotsu [Hi93, p. 194] for instance.) A special case of the spherical model is the intraclass correlation model such as

$$\boldsymbol{\Sigma}_i = \sigma_i^2\{(1 - \rho_i)\boldsymbol{I}_p + \rho_i \mathbf{1}\mathbf{1}'\}$$

for some ρ_i, where $\mathbf{1} = (1, ..., 1)'$. When $k = 1$ and the variance-covariance matrix has the spherical structure, Bhargava and Srivastava [BhS73], Chaturvedi et al. [CSS92], Hsu [Hs89], Nelson [Ne93], Nelson and Matejcik [NM95] and others considered problem (P5) by using an appropriate procedure. However, the case when $k \geq 2$ had not been tackled fully, considering interaction among populations.

When τ_i $(1 \leq i \leq k)$ is known, the sample sizes $\boldsymbol{n}$ needed to solve (P5) would be determined as the smallest integer such that $n_i \geq z^2|b_i|\tau_i \sum_{i=1}^k |b_i|\tau_i/d^2$ for each i if one considers minimizing the total sample size $\sum_{i=1}^k n_i$. Here, for each of (MCA), (MCB) and (MCC), $z > 0$ is determined as a solution to a certain equation depending on (p, α). (See Aoshima [A01] for the details.) When τ_i $(1 \leq i \leq k)$ is unknown, the sample sizes $\boldsymbol{n}$ should be determined through estimates of τ_i, $1 \leq i \leq k$. For each of (MCA), (MCB) and (MCC), problem (P5) is solved as follows.

[Two–stage procedure (P5)]

(T1) First, take a pilot sample $\boldsymbol{X}_{ij} = (X_{ij1}, ..., X_{ijp})'$, $j = 1, ..., m$, of size m (≥ 2) and calculate

$$S_{im}^2 = \nu^{-1} \sum_{r=1}^{p} \sum_{j=1}^{m} (X_{ijr} - \overline{X}_{ij.} - \overline{X}_{i.r} + \overline{X}_{i..})^2$$

with $\nu = (p-1)(m-1)$ as an estimate of τ_i^2 for each i. Here,

$$\overline{X}_{ij.} = p^{-1} \sum_{r=1}^{p} X_{ijr}, \qquad \overline{X}_{i.r} = m^{-1} \sum_{j=1}^{m} X_{ijr}$$

and

$$\overline{X}_{i..} = (pm)^{-1} \sum_{r=1}^{p} \sum_{j=1}^{m} X_{ijr}.$$

Define the stopping rule of each population by

$$N_i = \max\left\{m,\ \left[t_m^2 |b_i| S_{im} \sum_{i=1}^{k} |b_i| S_{im}/d^2\right] + 1\right\}, \tag{8}$$

where $t_m > 0$ is determined as a solution to the following equation for each of (MCA), (MCB) and (MCC):

$$\text{(MCA)}\quad kp\int_0^\infty\int_{-\infty}^\infty \left\{\Phi(x)-\Phi(x-t_m\sqrt{y/\nu})\right\}^{p-1}(1-F_\nu(y))^{k-1}\,d\Phi(x)dF_\nu(y) = 1-\alpha;$$

$$\text{(MCB)}\quad k\int_0^\infty\int_{-\infty}^\infty \left\{\Phi(x+t_m\sqrt{y/\nu})\right\}^{p-1}(1-F_\nu(y))^{k-1}\,d\Phi(x)dF_\nu(y) = 1-\alpha;$$

$$\text{(MCC)}\quad k\int_0^\infty\int_{-\infty}^\infty \left\{\Phi(x+t_m\sqrt{y/\nu})-\Phi(x-t_m\sqrt{y/\nu})\right\}^{p-1} \times (1-F_\nu(y))^{k-1}\,d\Phi(x)dF_\nu(y) = 1-\alpha,$$

where $\Phi(\cdot)$ denotes the standard normal distribution function and $F_r(\cdot)$ denotes the chi–squared distribution function with r d.f.

(T2) Next, take an additional sample $\boldsymbol{X}_{im+1}, ..., \boldsymbol{X}_{iN_i}$ of size $N_i - m$ for each i. Let $\boldsymbol{N} = (N_1, ..., N_k)$. By combining the initial sample and the additional sample, calculate $\overline{\boldsymbol{X}}_{iN_i} = N_i^{-1}\sum_{j=1}^{N_i}\boldsymbol{X}_{ij}$ for each i. Then, construct the simultaneous confidence intervals $SCI(\boldsymbol{T}_{\mathbf{N}})$ based on the components $(T_{1\mathbf{N}}, ..., T_{p\mathbf{N}})$ of $\boldsymbol{T}_{\mathbf{N}} = \sum_{i=1}^k b_i\overline{\boldsymbol{X}}_{iN_i}$.

Aoshima [A01] showed that the above two–stage procedure solves problem (P5), and also discussed the case when the $\boldsymbol{\Sigma}_i$'s do not have any specific structure. Let t_{ma}, t_{mb} and t_{mc} denote the t_m value for each of (MCA), (MCB) and (MCC). Then $t_{mb} < t_{mc} < t_{ma}$, and hence the sample should be required for each multiple comparisons method in such order. The reason why (MCB) requires the smallest sample size is because fewest confidence intervals are required to be simultaneously correct. Note that (MCB) implies the inference of both the indifference–zone and the subset selection methodologies for ranking and selection of the best component with a specified probability of correct selection. See Hsu [Hs96] for the details. A special case of (MCB) in this context was given by Hyakutake [H00] under intraclass correlation models when $k = 2$. For the second–order analysis related to procedure (P5), see Aoshima and Takada [AT00] and Aoshima and Miyajima [AMi01].

For other related topics to which two–stage procedures are applicable, Aoshima and Mukhopadhyay [AM98] considered Scheffé–type simultaneous confidence intervals for a double linear combination of $M = [\boldsymbol{\mu}_1, ..., \boldsymbol{\mu}_k]$. Hyakutake [H92] and Aoshima, Aoki and Kai [AAK04] considered an application to select a most preferable component of the Marshall-Olkin [MO67] multivariate exponential distribution in life testing. Aoshima and Chen [AC99] and Aoshima, Chen and Panchapakesan [ACP03] considered an application in a voter preference problem when there is a nuisance cell in which Dirichlet integrals, often used in inverse sampling, do not work. We note that the two–stage procedure is simpler and faster than a sequential procedure in making a survey of voters in which some voters change their minds quickly after they see news on TV and the Internet and it is essential to make the survey at one time point to as many voters as possible.

5. Beyond the conditions of the two–stage procedure

It is natural to enquire whether departure from normality has any adverse effect on the performance of the two–stage procedure. However, the robustness of the two–stage procedure had not been studied fully so far. When conditions (a) and (b), given in Section 2, are not satisfied, the distribution of a statistic is quite complicated and even an asymptotic expansion of the distribution is difficult to derive. Some robustness studies had been made by simulation for various types of departure from normality (see Ramkaran [Ra83] and its references) or had been done on some certain specific models (see Blumenthal and Govindarajulu [BG77] and Aoshima and Kano [AK97] for instance). To start this section, we shall theoretically attempt to show the robustness of the two–stage procedure along the lines of Aoshima and Wakaki [AW01].

Let $\boldsymbol{X}$, $\{\boldsymbol{X}_i;\ i \geq 1\}$ be a sequence of i.i.d. random vectors with values in R^p. Let $\boldsymbol{\mu} = E(\boldsymbol{X})$, $\boldsymbol{\Sigma} = Cov(\boldsymbol{X})$ $(> \boldsymbol{0})$, and let $\lambda \geq \lambda_2 \geq \ldots \geq \lambda_p$ be the latent roots of $\boldsymbol{\Sigma}$. Let X_j denote the j-th element of $\boldsymbol{X}$, and $\mu_{i_1 \cdots i_r}$ the moment of $\boldsymbol{X}$ defined by $\mu_{i_1 \cdots i_r} = E(X_{i_1} \cdots X_{i_r})$. Similarly, the corresponding cumulant of $\boldsymbol{X}$ is denoted by $\kappa_{i_1 \cdots i_r}$. Let K_r be the r-th order tensor whose $(i_1, \ldots, i_r)$-element is $\kappa_{i_1 \cdots i_r}$. Having recorded $\boldsymbol{X}_1, \ldots, \boldsymbol{X}_n$ of size n, $\boldsymbol{\mu}$ is estimated by the sample mean $\overline{\boldsymbol{X}}_n$. Let us consider the following inference problem about $\boldsymbol{\mu}$.

(P6) For a value $d > 0$ and a confidence coefficient $1 - \alpha > 0$ both given beforehand, determine the sample size n such that the ellipsoidal confidence region $ECR(\overline{\boldsymbol{X}}_n) = \{\boldsymbol{\mu} |\ n(\boldsymbol{\mu} - \overline{\boldsymbol{X}}_n)' \hat{\boldsymbol{\Sigma}}^{-1} (\boldsymbol{\mu} - \overline{\boldsymbol{X}}_n) \leq c\}$ for some $c > 0$ satisfies $P(\boldsymbol{\mu} \in ECR(\overline{\boldsymbol{X}}_n)) \geq 1 - \alpha$ for any $(\boldsymbol{\mu}, \boldsymbol{\Sigma})$ with the maximum diameter $\leq 2d$.

When the population distribution is $N_p(\boldsymbol{\mu}, \boldsymbol{\Sigma})$, we may use procedure (P2) given in Section 3 and define the region by $ECR(\overline{\boldsymbol{X}}_N) = \{\boldsymbol{\mu} |\ N(\boldsymbol{\mu} - \overline{\boldsymbol{X}}_N)' \boldsymbol{S}_m^{-1} (\boldsymbol{\mu} - \overline{\boldsymbol{X}}_N) \leq a_m\}$, where $\boldsymbol{S}_m$, a_m and N are given in (T1) of (P2) and $\overline{\boldsymbol{X}}_N$ is given in (T2) of (P2). Then, Healy [He56] showed that procedure (P2) gives a solution to problem (P6) with $P(\boldsymbol{\mu} \in ECR(\overline{\boldsymbol{X}}_N)) = 1 - \alpha$ for any $(\boldsymbol{\mu}, \boldsymbol{\Sigma})$. It is evident that the maximum diameter of $ECR(\overline{\boldsymbol{X}}_N) \leq 2d$ in view of the stopping rule (3). Note that the distribution of $T_N^2 = N(\overline{\boldsymbol{X}}_N - \boldsymbol{\mu})' \boldsymbol{S}_m^{-1} (\overline{\boldsymbol{X}}_N - \boldsymbol{\mu})$ is coincident with Hotelling's T^2 distribution, and hence the constant c is given by a_m, which is the upper α point of Hotelling's T^2 distribution.

It is interesting to investigate how robust the above solution is against departure from normality, that is, whether the region $ECR(\overline{\boldsymbol{X}}_N)$ given by a_m guarantees its confidence coefficient to be the required $1 - \alpha$. It is difficult to give the distribution of T_N^2 without the assumption of normality in a form that is easy to handle theoretically. We note that the distribution of Hotelling's T^2 statistic under nonnormality is quite difficult to derive even in a single–stage sampling scheme, for which its asymptotic expansion form was given by Kano [Ka95] and Fujikoshi [Fu97]. As for the two–stage sampling scheme, a different approach from those former studies is required for derivation of an asymptotic expansion of the distribution. Aoshima and Wakaki [AW01] gave an asymptotic expansion for the distribution of T_N^2 under nonnormality when $m \to \infty$ and investigated the robustness of the solution given by procedure (P2). We consider the following assumptions:

(A0) Cramér's condition for the joint distribution of $(\boldsymbol{X},\ \boldsymbol{X}\boldsymbol{X}')$ holds.

(A1) The maximum latent root λ of $\boldsymbol{\Sigma}$ is simple.

(A2) $E(||\boldsymbol{X}||^{8+r}) < \infty$, $r > 0$.

(A3) $\lim_{m \to \infty} md^2 = c_0$ for some constant $c_0 \in (0, a\lambda)$, where a denotes the upper α point of the chi–squared distribution with p d.f.

Then, the distribution of T_N^2 is given in an asymptotic expansion as follows:

$$P\left(T_N^2 \le x\right) = F_p(x) + m^{-1}\sum_{j=0}^{3}\beta_j F_{p+2j}(x) + o(m^{-1}),$$

where $F_r(\cdot)$ denotes a chi–squared distribution function with r d.f. Here, the coefficients $(\beta_0, ..., \beta_3)$ are determined depending on p, $\rho = (md^2)^{-1}a\lambda$ and nonnormality parameters such as

$$\kappa_3^{<1>} = \sum_{i=1}^{p}\sum_{j=1}^{p}\sum_{k=1}^{p}\kappa_{ijk}^2, \quad \kappa_3^{<2>} = \sum_{i=1}^{p}\sum_{j=1}^{p}\sum_{k=1}^{p}\kappa_{iij}\kappa_{jkk},$$

$$\eta_{31}^{<2>} = \sum_{j=1}^{p}\kappa_{11j}\kappa_{j11}, \quad \eta_{32}^{<2>} = \sum_{j=1}^{p}\sum_{k=1}^{p}\kappa_{11j}\kappa_{jkk},$$

$$\xi_3^{<2>} = \sum_{i=1}^{p}\sum_{j=2}^{p}\lambda_j(\lambda-\lambda_j)^{-1}\kappa_{1ij}^2, \quad \kappa_4^{<1>} = \sum_{i=1}^{p}\sum_{j=1}^{p}\kappa_{iijj}$$

and $\eta_{41}^{<1>} = \sum_{j=1}^{p}\kappa_{11jj}$. (See Aoshima and Wakaki [AW01] for the details.) When the population distribution is symmetric, $\beta_3 = 0$. When the population distribution is a normal distribution, the above formula is coincident with an asymptotic expansion of Hotelling's T^2 distribution. The average of T_N^2 depends on both the third- and fourth-order cumulants in the order of m^{-1}, while in a single–stage sampling scheme it depends only on the third-order cumulant. This reflects on various simulation studies conducted by Blumenthal and Govindarajulu [BG77], Ramkaran [Ra83] and others when $p = 1$. Putting $x = a_m$ in the above formula, one can asymptotically evaluate the coverage probability of the region $ECR(\overline{\boldsymbol{X}}_N)$ given by procedure (P2). We observe that if the population distribution is symmetric and longer-tailed than the normal distribution, the coverage probability exceeds $1-\alpha$ when $\rho < \rho^\star$ for some given $\rho^\star$ $(\in (1, 2))$ depending on p. Further, if one applies Cornish-Fisher expansion to the distribution such as $P(T_N^2 \le a + bm^{-1}) = 1 - \alpha + o(m^{-1})$, we could obtain the constant $a + bm^{-1}$ as an option to modify procedure (P2) with a_m against nonnormality. After taking this modification, we observe in the situation described above that the average sample size required in procedure (P2) becomes less than that required in constructing the region $ECR(\overline{\boldsymbol{X}}_n) = \{\boldsymbol{\mu} | \ n(\boldsymbol{\mu} - \overline{\boldsymbol{X}}_n)'\boldsymbol{\Sigma}^{-1}(\boldsymbol{\mu} - \overline{\boldsymbol{X}}_n) \le a\}$ with confidence $1 - \alpha + o(m^{-1})$ for known $\boldsymbol{\Sigma}$. It would be interesting to consider an improvement of the estimate with higher order moments as seen in Uno and Isogai [UI00]. As for an improvement of the stopping rule, a monotone Bartlett–type correction for the T^2 statistic might be considered under nonnormality by extending the techniques of Fujikoshi [Fu00] and others.

Let us consider the estimation of parameters other than the mean. For instance, when considering a bounded risk problem about the variance of a normal distribution, the conditions on independence of estimates required for the two–stage procedure are not satisfied. Birnbaum and Healy [BH60] proposed a different two–stage procedure in which the ultimate estimator is defined by using only an additional sample. This procedure yields a bounded risk point estimate not only for the variance of a normal distribution but also for the parameters of a Poisson distribution, a binomial distribution and a hypergeometric distribution, and also for a scale parameter of the location–scale family of distributions: If followed by the use of Tchebychev's inequality, a fixed–width confidence interval of given confidence is naturally produced. Blum and Rosenblatt [BR69b] applied a similar technique to obtain a fixed–width confidence interval of given confidence for the moments of a distribution with increasing failure rate. However, the procedure given by [BH60] necessarily causes inefficiency, so it was necesssary to develop techniques so as to overcome this inconvenience. For estimation for the variance of a normal distribution and for the parameter of a uniform distribution $U(0,\theta)$, Graybill and Connell [GC64ab] gave a technique to reduce the sample size by using an inequality inherent to those distributions instead of Tchebychev's inequality. (That technique was applied by Takada [T86], Aoshima and Govindarajulu [AG02], and others.) For point estimation with bounded risk for a parameter of the scale family of distributions, Kubokawa [K89] proposed a technique to improve the ultimate estimator by combining with an initial sample. Later, Kubokawa [K90] applied this technique to point estimation with bounded risk for the generalized variance $|\boldsymbol{\Sigma}|$ of $N_p(\boldsymbol{\mu}, \boldsymbol{\Sigma})$. In addition, especially with fixed–width confidence intervals for the variance, Sproule [Sp74] applied an appropriate large sample theory of U-statistics to the sample variance in a purely sequential sampling scheme. See Ghosh et al. [GMS97] for related references.

In this article, several topics have been omitted for brevity, especially about other loss functions such as those based on an asymmetric loss or a squared loss modified at the boundary of the parameter space. The loss function used in sequential analysis does not have the invariant property of scale transformations as seen in this article. This is because the sample size is determined by picking up information about the population distribution through estimation of the scale parameter. There is very little work on estimation for nonlinear functions of means using two–stage procedures. Recently, Zheng et al. [ZSS98ab] considered a two–stage procedure for the problem of estimating the product of means, which arises in situations of determining area based on measurements of length and width in environmental applications. There are other problems, such as sequential time series (Sriram [Srir87,01], Lee and Sriram [LS99], Galtchouk and Konev [GK01], Shiohama and Taniguchi [ST01], and others), sequential change–point detection (Lai [La95], Siegmund and Venkatraman [SV95], Yakir [Y98], and others), and sequential density estimation (Isogai [I93, 99], Martinsek [Ma92, 93], Xu and Martinsek [XM95], Honda [Hon98], and others), when considering the precision of inference beforehand. However, judging from the character of the research fields, these problems should be handled with sequential sampling schemes rather than two–stage sampling schemes. Lastly, it would be interesting if a necessary and sufficient number of stages for sampling could be determined in some sense, and it might be applied to computer simulations and engineering in the near future.

References

[AkK96] Akahira, M. and Koike, K., On some properties of statistical sequential decision rules (in Japanese), Sugaku **48** (1996), 184–195; English transl., Sugaku Expositions **11** (1998), 197–213.

[Alb92] Albers, W., Asymptotic expansions for two-stage rank tests, Ann. Inst. Statist. Math., **44** (1992), 335–356.

[Alv77] Alvo, M., Bayesian sequential estimation, Ann. Statist., **5** (1977), 955–968.

[Alv78] Alvo, M., Sequential estimation of a truncation parameter, J. Amer. Statist. Assoc., **73** (1978), 404–407.

[A94a] Aoshima, M., Healy's sample size of two–stage procedure in heteroscedastic simultaneous inference, Commun. Statist.—Theory Meth., **23** (1994), 1297–1310.

[A94b] Aoshima, M., In heteroscedastic simultaneous inference, the Heteroscedastic Method II is more economical than the original HM, Amer. J. Math. Manage. Sci., **14** (1994), 167–196.

[A98] Aoshima, M., A two–stage procedure for fixed–size confidence region when covariance matrices have some structures, J. Japan Statist. Soc., **28** (1998), 59–67.

[A00] Aoshima, M., Second–order properties of improved two–stage procedure for a multivariate normal distribution, Commun. Statist.—Theory Meth., **29** (2000), 611–622.

[A01] Aoshima, M., Sample size determination for multiple comparisons with components of a linear function of mean vectors, Statistics: Reflections on the Past and Visions for the Future: An International Conference in Honor of Prof. C.R. Rao on the Occasion of His 80th Birthday, R. Khattree and S.D. Peddada, eds., Commun. Statist.—Theory Meth., **30** (2001), 1773–1788.

[AAK04] Aoshima, M., Aoki, M. and Kai, M., Two–stage procedures for selecting the best component of a multivariate exponential distribution, Applied Sequential Methodologies, N. Mukhopadhyay, S. Datta and S. Chattopadhyay, eds., Marcel Dekker, Inc., New York, 2004, 19–34.

[AC99] Aoshima, M. and Chen, P., A two–stage procedure for selecting the largest multinomial cell probability when nuisance cell is present, Sequential Anal., **18** (1999), 143–155.

[ACP03] Aoshima, M., Chen, P. and Panchapakesan, S., Sequential procedures for selecting the most probable multinomial cell when a nuisance cell is present, Commun. Statist.—Theory Meth., **32** (2003), 893–906.

[AG02] Aoshima, M. and Govindarajulu, Z., Fixed–width confidence interval for a lognormal mean, Internat. J. Math. Math. Sci., **29** (2002), 143–154.

[AK97] Aoshima, M. and Kano, Y., A note on robustness of two–stage procedure for a multivariate compounded normal distribution, Sequential Anal., **16** (1997), 175–187.

[AMi01] Aoshima, M. and Miyajima, H., Sample size reduction of two–stage procedure, Kokyuroku 1224, RIMS, Kyoto, 2001, 61–72.

[AM98] Aoshima, M. and Mukhopadhyay, N., Fixed–width simultaneous confidence intervals for multinormal means in several intraclass correlation models, J. Multiv. Anal., **66** (1998), 46–63.

[AM99] Aoshima, M. and Mukhopadhyay, N., Second–order properties of a two–stage fixed–size confidence region when the covariance matrix has a structure, Statistical Inference and Data Analysis–A Meeting in Honour of Prof. N. Sugiura, Commun. Statist.—Theory Meth., **28** (1999), 839–855.

[AM02] Aoshima, M. and Mukhopadhyay, N., Two–stage estimation of a linear function of normal means with second–order approximations, Sequential Anal., **21** (2002), 109–144.

[AT00] Aoshima, M. and Takada, Y., Second–order properties of a two–stage procedure for comparing several treatments with a control, J. Japan Statist. Soc., **30** (2000), 27–41.

[AT02] Aoshima, M. and Takada, Y., Bounded risk point estimation of a linear function of k multinormal mean vectors when covariance matrices are unknown, Advances on Theoretical and Methodological Aspects of Probability and Statistics, N. Balakrishnan, ed., Taylor & Francis, New York, 2002, 279–287.

[ATS02] Aoshima, M., Takada, Y. and Srivastava, M.S., A two–stage procedure for estimating a linear function of k multinormal mean vectors when covariance matrices are unknown, J. Statist. Plan. Inference, **100** (2002), 109–119.

[AW01] Aoshima, M. and Wakaki, H., Two–stage procedure under nonnormality, Bull. International Statist. Inst. 53rd Session, **1** (2001), 99–100.

[ABG49] Arrow, K.J., Blackwell, D. and Girshick, M.A., Bayes and minimax solutions of sequential decision problems, Econometrica, **17** (1949), 213–244.

[BaS56] Bahadur, R.R. and Savage, L.J., The nonexistence of certain statistical procedures in nonparametric problems, Ann. Math. Statist., **27** (1956), 1115–1122.

[B67] Banerjee, S., Confidence interval of preassigned length for the Behrens–Fisher problem, Ann. Math. Statist., **38** (1967), 1175–1179.

[BhS73] Bhargava, R.P. and Srivastava, M.S., On Tukey's confidence intervals for the contrasts in the means of the intraclass correlation model, J. Roy. Statist. Soc., **B 35** (1973), 147–152.

[BY68] Bickel, P.J. and Yahav, J.A., Asymptotically optimal Bayes and minimax procedures in sequential estimation, Ann. Math. Statist., **39** (1968), 442–456.

[BH60] Birnbaum, A., and Healy, W.C., Jr., Estimates with prescribed variance based on two–stage sampling, Ann. Math. Statist., **31** (1960), 662–676.

[BR63] Blum, J.R. and Rosenblatt, J., On multistage estimation, Ann. Math. Statist., **34** (1963), 1452–1458.

[BR66] Blum, J.R. and Rosenblatt, J., On some statistical problems requiring purely sequential sampling schemes, Ann. Inst. Statist. Math., **18** (1966), 351–355.

[BR69a] Blum, J.R. and Rosenblatt, J., On fixed precision estimation in time series, Ann. Math. Statist., **40** (1969), 1021–1032.

[BR69b] Blum, J.R. and Rosenblatt, J., Fixed precision estimation in the class of IFR distributions, Ann. Inst. Statist. Math., **40** (1969), 211–213.

[BG77] Blumenthal, S. and Govindarajulu, Z., Robustness of Stein's two–stage procedure for mixtures of normal populations, J. Amer. Statist. Assoc., **72** (1977), 192–196.

[Cha50] Chapman, D.G., Some two sample tests, Ann. Math. Statist., **21** (1950), 601–606.

[C59] Chatterjee, S.K., On an extension of Stein's two–sample procedure to the multinormal problem, Culcutta Statist. Assoc. Bull., **8** (1959), 121–148.

[C60] Chatterjee, S.K., Some further results on the multinormal extension of Stein's two–sample procedure, Culcutta Statist. Assoc. Bull., **9** (1960), 20–28.

[C91] Chatterjee, S.K., Two–stage and multistage procedures, Handbook of Sequential Analysis, B.K. Ghosh and P.K. Sen, eds., Marcel Dekker, Inc., New York, 1991, 21–45.

[CSS92] Chaturvedi, A., Shukla, N.D. and Shukla, P.S., Fixed–width confidence intervals for contrasts in the means, Ann. Inst. Statist. Math., **44** (1992), 157–167.

[CR65] Chow, Y.S. and Robbins, H., On the asymptotic theory of fixed–width sequential confidence intervals for the mean, Ann. Math. Statist., **36** (1965), 457–462.

[CS84a] Cohen, A. and Sackrowitz, H.B., Bayes double sample estimation procedures, Ann. Statist., **12** (1984), 1035–1049.

[CS84b] Cohen, A. and Sackrowitz, H.B., Results on double sample estimation for the binomial distribution, Ann. Statist., **12** (1984), 1109–1116.

[D40] Dantzig, G.B., On the non-existence of tests of "Student's" hypothesis having power functions independent of σ, Ann. Math. Statist., **11** (1940), 186–192.

[DG00] Dmitrienko, A. and Govindarajulu, Z., Sequential confidence regions for maximum likelihood estimates, Ann. Statist., **28** (2000), 1472–1501.

[DB79] Dudewicz, E.J. and Bishop, T.A., The heteroscedastic method, Optimizing Methods in Statistics (Proc. Internat. Conf., Bombay, 1977; dedicated to H. Scheffé), J.S. Rustagi, ed., Academic Press, New York, 1979, 183–203.

[DT87] Dudewicz, E.J. and Taneja, V.S., The Heteroscedastic Method: Multivariate implementation, Ann. Inst. Statist. Math., **39** (1987), 177–190.

[F66] Farrell, R.H., Bounded length confidence intervals for the $p-$point of a distribution function, III, Ann. Math. Statist., **37** (1966), 586–592.

[Fu97] Fujikoshi, Y., An asymptotic expansion for the distribution of Hotelling's T^2-statistic under nonnormality, J. Multiv. Anal., **61** (1997), 187–193.

[Fu00] Fujikoshi, Y., Transformations with improved chi–squared approximations, J. Multiv. Anal., **72** (2000), 249–263.

[GK01] Galtchouk, L. and Konev, V., On sequential estimation of parameters in semimartingale regression models with continuous time parameter, Ann. Statist., **29** (2001), 1508–1536.

[Gh75] Ghosh, B.K., A two–stage procedure for the Behrens-Fisher problem, J. Amer. Statist. Assoc., **70** (1975), 457–462.

[GMS97] Ghosh, M., Mukhopadhyay, N. and Sen, P.K., Sequential Estimation, Wiley, New York, 1997.

[GNS87] Ghosh, M., Nickerson, D.M. and Sen, P.K., Sequential shrinkage estimation, Ann. Statist., **15** (1987), 817–829.

[GS83] Ghosh, M. and Sen, P.K., On two–stage James–Stein estimators, Sequential Anal., **2** (1983), 359–367.

[G58] Ghurye, S.G., Note on sufficient statistics and two-stage procedures, Ann. Math. Statist., **29** (1958), 155–166.

[GC64a] Graybill, F.A. and Connell, T.L., Sample size required for estimating the variance within d units of the true value, Ann. Math. Statist., **35** (1964), 438–440.

[GC64b] Graybill, F.A. and Connell, T.L., Sample size required to estimate the parameter in the uniform density within d units of the true value, J. Amer. Statist. Assoc., **59** (1964), 550–556.

[Ha81] Hall, P., Asymptotic theory of triple sampling for sequential estimation of a mean, Ann. Statist., **9** (1981), 1229–1238.

[Ha83] Hall, P., Sequential estimation saving sampling operations, J. Roy. Statist. Soc., **B 45** (1983), 219–223.

[He56] Healy, W.C., Jr., Two–sample procedures in simultaneous estimation, Ann. Math. Statist., **27** (1956), 687–702.

[HeS83] Hewett, J.E. and Spurrier, J.D., A survey of two stage tests of hypotheses: Theory and applications, Commun. Statist.-Theory Meth., **12** (1983), 2307–2425.

[Hi92] Hirotsu, C., Analysis for Experimental Data: Beyond Analysis of Variance (in Japanese), Kyoritsu, 1992.

[Hi93] Hirotsu, C., Beyond analysis of variance techniques: Some applications in clinical trials, Int. Statist. Rev., **61** (1993), 183–201.

[HT87] Hochberg Y. and Tamhane, A.C., Multiple Comparison Procedures, John Wiley and Sons, Inc., New York, 1987.

[Hol95] Holm, S., Confidence sets of fixed size with predetermined confidence level, Commun. Statist.—Theory Meth., **24** (1995), 1521–1536.

[Hol99] Holm, S., Confidence sets of fixed size for multinormal means, J. Statist. Plan. Inference, **82** (1999), 229–236.

[Hon98] Honda, T., Sequential estimation of the marginal density function for a strongly mixing process, Sequential Anal., **17** (1998), 239–251.

[Hs89] Hsu, J.C., Sample size computation for designing multiple comparison experiments, Comp. Statist. & Data Anal., **7** (1989), 79–91.

[Hs96] Hsu, J.C., Multiple Comparisons: Theory and Methods, Chapman & Hall, Inc., New York, 1996.

[HC82] Hwang, J.T. and Casella, G., Minimax confidence sets for the mean of a multivariate normal distribution, Ann. Statist., **10** (1982), 868–881.

[HC84] Hwang, J.T. and Casella, G., Improved set estimators for a multivariate normal mean, Statist. Decisions Suppl., **1** (1984), 3–16.

[H86] Hyakutake, H., A construction method of certain matrices required in the multivariate heteroscedastic method, Ann. Inst. Statist. Math., **38** (1986), 523–528.

[H92] Hyakutake, H., Selecting the better component of a bivariate exponential distribution, Statist. Decisions, **10** (1992), 153–162.

[H98] Hyakutake, H., Fixed–size confidence region of the difference of two multinormal means when the covariance matrices have structure, J. Japan Statist. Soc., **28** (1998), 175–180.

[H00] Hyakutake, H., Selecting the component with maximal difference of two multinormal means, Amer. J. Math. Manage. Sci., **20** (2000), 433–442.

[HS87] Hyakutake, H. and Siotani, M., The multivariate heteroscedastic method: Distributions of statistics and an application, Amer. J. Math. Manage. Sci., **7** (1987), 89–111.

[HS89] Hyakutake, H. and Siotani, M., Mean and variance of sample size in multivariate heteroscedastic method, Contributions to Probability and Statistics: Essays in Honor of Ingram Olkin, Springer-Verlag, New York, 1989, 227–240.

[HTA95] Hyakutake, H., Takada, Y. and Aoshima, M., Fixed–size confidence regions for the multinormal mean in an intraclass correlation model, Amer. J. Math. Manage. Sci., **15** (1995), 291–308.

[I93] Isogai, E., Recursive estimation of non-parametric probability density functions (in Japanese), Sugaku, **45** (1993), 27–41; English transl., Amer. Math. Soc. Transl. (2) **161** (1994), 135–151.

[I99] Isogai, E., Sequential estimation of a density and its derivatives with bounded MISE, J. Statist. Plan. Inference, **79** (1999), 59–78.

[Ka95] Kano, Y., An asymptotic expansion of the distribution of Hotelling's T^2-statistic under general condition, Amer. J. Math. Manage. Sci., **15** (1995), 317–341.

[Ko99] Koike, K., A lower bound for the Bayes risk in the sequential case, Commun. Statist.—Theory Meth., **28** (1999), 857–871.

[KOR64] Koopmans, L.H., Owen, D.B. and Rosenblatt, J.I., Confidence intervals for the coefficient of variation for the normal and log normal distributions, Biometrika, **51** (1964), 25–32.

[K89] Kubokawa, T., Improving on two–stage estimators for scale families, Metrika, **36** (1989), 7–13.

[K90] Kubokawa, T., Two–stage estimators with bounded risk in a growth curve model, J. Japan Statist. Soc., **20** (1990), 77–87.

[KS94] Kubokawa, T. and Saleh, A.K.Md.E., Two–stage point estimation with a shrinkage stopping rule, Metrika, **41** (1994), 293–306.

[La95] Lai, T.L., Sequential change-point detection in quality control and dynamical systems, J. Roy. Statist. Soc., **B 57** (1995), 613–658.

[LS99] Lee, S. and Sriram, T.N., Sequential point estimation of parameters in a threshold AR(1) model, Stochastic Process. Appl., **84** (1999), 343–355.

[L51] Lehmann, E.L., Notes on the Theory of Estimation, Univ. of California Press, Berkeley, 1951.

[Li97a] Liu, W., Improving the fully sequential sampling scheme of Anscombe-Chow-Robbins, Ann. Statist., **25** (1997), 2164–2171.

[Li97b] Liu, W., A k-stage sequential sampling procedure for estimation of normal mean, J. Statist. Plan. Inference, **65** (1997), 109–127.

[Li97c] Liu, W., On some sample size formulae for controlling both size and power in clinical trials, The Statistician, **46** (1997), 239–251.

[MO67] Marshall, A.W. and Olkin, I., A multivariate exponential distribution, J. Amer. Statist. Assoc., **62** (1967), 30–44.

[Ma87] Martinsek, A.T., Empirical Bayes methods in sequential estimation, Sequential Anal., **6** (1987), 119–137.

[Ma92] Martinsek, A.T., Using stopping rules to bound the mean integrated squared error in density estimation, Ann. Statist., **20** (1992), 797–806.

[Ma93] Martinsek, A.T., Fixed width confidence bands for density functions, Sequential Anal., **12** (1993), 169–177.

[MF93] Mukaihata, S. and Fujikoshi, Y., Error bounds for asymptotic expansions of some distributions in a multivariate two–stage procedure, Hiroshima Math. J., **23** (1993), 565–575.

[M82] Mukhopadhyay, N., Stein's two-stage procedure and exact consistency, Scand. Actuarial J., **1982**, 110–122.

[M99] Mukhopadhyay, N., Second–order properties of a two–stage fixed–size confidence region for the mean vector of a multivariate normal distribution, J. Multiv. Anal., **68** (1999), 250–263.

[MD97] Mukhopadhyay, N. and Duggan, W.T., Can a two–stage procedure enjoy second–order properties?, Sankhya, Series A, **59** (1997), 435–448.

[MD99] Mukhopadhyay, N. and Duggan, W.T., On a two–stage procedure having second–order properties with applications, Ann. Inst. Statist. Math., **51** (1999), 621–636.

[ML89] Mukhopadhyay, N. and Liberman, S., Sequential estimation of a linear function of mean vectors, Sequential Anal., **8** (1989), 381–395.

[N96] Nagao, H., On fixed width confidence regions for multivariate normal mean when the covariance matrix has some structure, Sequential Anal., **15** (1996), 37–46.

[N97a] Nagao, H., Asymptotically pointwise optimal rules for estimating the mean in general exponential distributions for squared loss, Sequential Anal., **16** (1997), 155–174.

[N97b] Nagao, H., Asymptotically pointwise optimal rules of sequential estimation of mean vector when an information matrix has some structure in a multivariate normal population, Sequential Anal., **16** (1997), 363–374.

[NY97] Nagata, Y. and Yoshida, M., Fundamentals of Statistical Multiple Comparisons (in Japanese), Scientist, 1997.

[NS85] Natarajan, J. and Strawderman, W.E., Two–stage sequential estimation of a multivariate normal mean under quadratic loss, Ann. Statist., **13** (1985), 1509–1522.

[Ne93] Nelson, B.L., Robust multiple comparisons under common tandom numbers, ACM Trans. Model. Comp. Simu., **3** (1993), 225–243.

[NM95] Nelson, B.L. and Matejcik, F.J., Using common random numbers for indifference–zone selection and multiple comparisons in simulation, Manage. Sci., **41** (1995), 1935–1945.

[R83] Ramkaran, The robustness of Stein's two–stage procedure, Ann. Statist., **11** (1983), 1251–1256.

[R73] Rao, C.R., Linear Statistical Inference and Its Application, 2nd ed., Wiley, New York, 1973.

[Sc93] Schwabe, R., Some two–stage procedures for treating the Behrens–Fisher problem, Model-Oriented Data Analysis, W.G. Müller, H. P. Wynn and A. A. Zhigljavsky, eds., Physica, Heidelberg (1993), 81–89.

[Sh89] Shinozaki, N., Improved confidence sets for the mean of a multivariate normal distribution, Ann. Inst. Statist. Math., **41** (1989), 331–346.

[ST01] Shiohama, T. and Taniguchi, M., Sequential estimation for a functional of the spectral density of a Gaussian stationary process, Ann. Inst. Statist. Math., **53** (2001), 142–158.

[SV95] Siegmund, D. and Venkatraman, E.S., Using the generalized likelihood ratio statistics for sequential detection of a change-point, Ann. Statist., **23** (1995), 255–271.

[Sin63] Singh, R., Existence of bounded length confidence intervals, Ann. Math. Statist., **34** (1963), 1474–1485.

[Sio87] Siotani, M., Multivariate Behrens–Fisher problem by heteroscedastic method, Advances in Multivariate Statistical Analysis, A.K. Gupta, ed., Reidel, Dordrecht, 1987, 327–340.

[SHF85] Siotani, M., Hayakawa, T. and Fujikoshi, Y., Modern Multivariate Statistical Analysis: A Graduate Course and Handbook, American Sciences Press, Inc., Columbus, Ohio, 1985.

[Sp74] Sproule, R.N., Asymptotic properties of U-statistics, Trans. Amer. Math. Soc., **199** (1974), 55–64.

[Srir87] Sriram, T.N., Sequential estimation of the mean of a first–order stationary autoregressive process, Ann. Statist., **15** (1987), 1079–1090.

[Srir01] Sriram, T.N., Fixed size confidence regions for parameters of threshold AR(1) models, J. Statist. Plan. Inference, **97** (2001), 293–304.

[Sriv67] Srivastava, M.S., On fixed–width confidence bounds for regression parameters and the mean vector, J. Roy. Statist. Soc., **B 29** (1967), 132–140.

[S45] Stein, C., A two–sample test for a linear hypothesis whose power is independent of the variance, Ann. Math. Statist., **16** (1945), 243–258.

[T86] Takada, Y., Non–existence of fixed sample size procedures for scale families, Sequential Anal., **5** (1986), 93–101.

[T98a] Takada, Y., The nonexistence of procedures with bounded performance characteristics in certain parametric inference problems, Ann. Inst. Statist. Math., **50** (1998), 325–335.

[T98b] Takada, Y., Asymptotic improvement of the usual confidence set in a multivariate normal distribution with unknown variance, J. Multiv. Anal., **64** (1998), 118–130.

[TA96] Takada, Y. and Aoshima, M., Two–stage procedures for the difference of two multinormal means with covariance matrices different only by unknown scalar multipliers, Commun. Statist.—Theory Meth., **25** (1996), 2371–2379.

[TA97] Takada, Y. and Aoshima, M., Two–stage procedures for estimating a linear function of multinormal mean vectors, Sequential Anal., **16** (1997), 353–362.

[TH97] Takada, Y. and Hyakutake, H., Fixed–size confidence region by multivariate two–stage procedure when the covariance matrix has a structure, J. Japan Statist. Soc., **27** (1997), 37–44.

[UI00] Uno, C. and Isogai, E., Sequential estimation of a linear combination of means, Statist. Probab. Lett., **47** (2000), 45–52.

[W77] Woodroofe, M., Second–order approximations for sequential point and interval estimation, Ann. Statist., **5** (1977), 984–995.

[W81] Woodroofe, M., A.P.O. rules are asymptotically nondeficient for estimation with squared error loss, Z. Wahrschein. Verw. Geb., **58** (1981), 331–341.

[XM95] Xu, Y. and Martinsek, A.T., Sequential confidence bands for densities, Ann. Statist., **23** (1995), 2218–2240.

[Y98] Yakir, B., On the average run length to false alarm in surveillance problems which possess an invariance structure, Ann. Statist., **26** (1998), 1198–1214.

[Z71] Zacks, S., The Theory of Statistical Inference, Wiley, New York, 1971.

[ZSS98a] Zheng, S., Seila, A.F. and Sriram, T.N., Sequential fixed–width confidence interval for the product of two means, Ann. Inst. Statist. Math., **50** (1998), 119–145.

[ZSS98b] Zheng, S., Seila, A.F. and Sriram, T.N., Asymptotically risk efficient two stage procedure for estimating the product of k (≥ 2) means, Statist. Decisions, **16** (1998), 369–387.

INSTITUTE OF MATHEMATICS, UNIVERSITY OF TSUKUBA, IBARAKI 305-8571, JAPAN
E-mail address: aoshima@@math.tsukuba.ac.jp

Translated by MAKOTO AOSHIMA

Titles in This Series

215 Selected Papers on Differential Equations and Analysis

214 **N. N. Uraltseva, Editor,** Proceedings of the St. Petersburg Mathematical Society, Volume X

213 **Ernest Vinberg, Editor,** Lie Groups and Invariant Theory

212 **V. M. Buchstaber and I. M. Krichever, Editors,** Geometry, Topology, and Mathematical Physics

211 **K. Nomizu, Editor,** Selected Papers on Analysis and Differential Equations

210 **S. G. Gindikin, Editor,** Lie Groups and Symmetric Spaces. In Memory of F. I. Karpelevich

209 **N. N. Uraltseva, Editor,** Proceedings of the St. Petersburg Mathematical Society, Volume IX

208 **M. V. Karasev, Editor,** Asymptotic Methods for Wave and Quantum Problems

207 **Yu. M. Suhov, Editor,** Analytic Methods in Applied Probability

206 **M. A. Shubin and M. S. Agranovich, Editors,** Partial Differential Equations (Mark Viskik's Seminar)

205 **N. N. Uraltseva, Editor,** Proceedings of the St. Petersburg Mathematical Society, Volume VIII

204 **K. Nomizu, Editor,** Selected Papers on Classical Analysis

203 **N. N. Uraltseva, Editor,** Proceedings of the St. Petersburg Mathematical Society, Volume VII

202 **V. Turaev and A. Vershik, Editors,** Topology, Ergodic Theory, Real Algebraic Geometry. Rokhlin's Memorial

201 **Michael Semenov-Tian-Shansky, Editor,** L. D. Faddeev's Seminar on Mathematical Physics

200 **L. Lerman, G. Polotovskiĭ, and L. Shilnikov, Editors,** Methods of Qualitative Theory of Differential Equations and Related Topics

199 **N. N. Uraltseva, Editor,** Proceedings of the St. Petersburg Mathematical Society, Volume VI

198 **R. A. Minlos, Senya Shlosman, and Yu. M. Suhov, Editors,** On Dobrushin's Way. From Probability Theory to Statistical Physics

197 **Vladimir Arnold, Maxim Kontsevich, and Anton Zorich, Editors,** Pseudoperiodic Topology

196 **Ya. Eliashberg, D. Fuchs, T. Ratiu, and A. Weinstein, Editors,** Northern California Symplectic Geometry Seminar

195 **O. V. Belegradek et al.,** Model Theory and Applications

194 **A. Astashkevich and S. Tabachnikov, Editors,** Differential Topology, Infinite-Dimensional Lie Algebras, and Applications (D. B. Fuchs' 60th Anniversary Collection)

193 **N. N. Uraltseva, Editor,** Proceedings of the St. Petersburg Mathematical Society, Volume V

192 **Lev Beklemishev, Mati Pentus, and Nikolai Vereshchagin,** Provability, Complexity, Grammars

191 **A. Yu. Morozov and M. A. Olshanetsky, Editors,** Moscow Seminar in Mathematical Physics

190 **S. Tabachnikov, Editor,** Differential and Symplectic Topology of Knots and Curves

189 **V. Buslaev, M. Solomyak, and D. Yafaev, Editors,** Differential Operators and Spectral Theory (M. Sh. Birman's 70th Anniversary Collection)

188 **O. A. Ladyzhenskaya, Editor,** Proceedings of the St. Petersburg Mathematical Society, Volume IV

TITLES IN THIS SERIES

187 **M. V. Karasev, Editor,** Coherent Transform, Quantization, and Poisson Geometry

186 **A. Khovanskiĭ, A. Varchenko, and V. Vassiliev, Editors,** Geometry of Differential Equations

185 **B. Feigin and V. Vassiliev, Editors,** Topics in Quantum Groups and Finite-Type Invariants (Mathematics at the Independent University of Moscow)

184 **Peter Kuchment and Vladimir Lin, Editors,** Voronezh Winter Mathematical Schools (Dedicated to Selim Krein)

183 **K. Nomizu, Editor,** Selected Papers on Harmonic Analysis, Groups, and Invariants

182 **V. E. Zakharov, Editor,** Nonlinear Waves and Weak Turbulence

181 **G. I. Olshanski, Editor,** Kirillov's Seminar on Representation Theory

180 **A. Khovanskiĭ, A. Varchenko, and V. Vassiliev, Editors,** Topics in Singularity Theory

179 **V. M. Buchstaber and S. P. Novikov, Editors,** Solitons, Geometry, and Topology: On the Crossroad

178 **V. Kreinovich and G. Mints, Editors,** Problems of Reducing the Exhaustive Search

177 **R. L. Dobrushin, R. A. Minlos, M. A. Shubin, and A. M. Vershik, Editors,** Topics in Statistical and Theoretical Physics (F. A. Berezin Memorial Volume)

176 **E. V. Shikin, Editor,** Some Questions of Differential Geometry in the Large

175 **R. L. Dobrushin, R. A. Minlos, M. A. Shubin, and A. M. Vershik, Editors,** Contemporary Mathematical Physics (F. A. Berezin Memorial Volume)

174 **A. A. Bolibruch, A. S. Merkur′ev, and N. Yu. Netsvetaev, Editors,** Mathematics in St. Petersburg

173 **V. Kharlamov, A. Korchagin, G. Polotovskiĭ, and O. Viro, Editors,** Topology of Real Algebraic Varieties and Related Topics

172 **K. Nomizu, Editor,** Selected Papers on Number Theory and Algebraic Geometry

171 **L. A. Bunimovich, B. M. Gurevich, and Ya. B. Pesin, Editors,** Sinai's Moscow Seminar on Dynamical Systems

170 **S. P. Novikov, Editor,** Topics in Topology and Mathematical Physics

169 **S. G. Gindikin and E. B. Vinberg, Editors,** Lie Groups and Lie Algebras: E. B. Dynkin's Seminar

168 **V. V. Kozlov, Editor,** Dynamical Systems in Classical Mechanics

167 **V. V. Lychagin, Editor,** The Interplay between Differential Geometry and Differential Equations

166 **O. A. Ladyzhenskaya, Editor,** Proceedings of the St. Petersburg Mathematical Society, Volume III

165 **Yu. Ilyashenko and S. Yakovenko, Editors,** Concerning the Hilbert 16th Problem

164 **N. N. Uraltseva, Editor,** Nonlinear Evolution Equations

163 **L. A. Bokut′, M. Hazewinkel, and Yu. G. Reshetnyak, Editors,** Third Siberian School "Algebra and Analysis"

162 **S. G. Gindikin, Editor,** Applied Problems of Radon Transform

161 **K. Nomizu, Editor,** Selected Papers on Analysis, Probability, and Statistics

160 **K. Nomizu, Editor,** Selected Papers on Number Theory, Algebraic Geometry, and Differential Geometry

159 **O. A. Ladyzhenskaya, Editor,** Proceedings of the St. Petersburg Mathematical Society, Volume II

For a complete list of titles in this series, visit the AMS Bookstore at **www.ams.org/bookstore/**.